Systematics, Ecology, and Chromosomal Evolution of *Melampodium* (Asteraceae)

Melampodium divaricatum, an attractive species widely distributed throughout Latin America.

Tod F. Stuessy • Hanna Weiss-Schneeweiss
Gerald M. Schneeweiss • José L. Villaseñor

Systematics, Ecology, and Chromosomal Evolution of *Melampodium* (Asteraceae)

Analysis of Species from Latin America and the Southwestern United States

Springer

Tod F. Stuessy
Department of Evolution,
Ecology & Organismal Biology
The Ohio State University
Columbus, OH, USA

Gerald M. Schneeweiss
Department of Botany & Biodiversity
Research
University of Vienna
Vienna, Austria

Hanna Weiss-Schneeweiss
Department of Botany & Biodiversity
Research
University of Vienna
Vienna, Austria

José L. Villaseñor
Instituto de Biología
Universidad Nacional Autónoma de México
Ciudad de México, México

ISBN 978-3-031-81994-0 ISBN 978-3-031-81995-7 (eBook)
https://doi.org/10.1007/978-3-031-81995-7

This Springer imprint is published by the registered company Springer Nature Switzerland AG
The registered company address is: Gewerbestrasse 11, 6330 Cham, Switzerland

We dedicate this book to the memory of Billie Lee Turner (1925–2020), former Professor at the University of Texas, Austin, who was responsible for opening the door to Melampodium *for the senior author, which eventually made this book possible. He was an outstanding plant systematist and a stimulating and supportive mentor.*

Preface

How and why we have dedicated time and resources to understanding *Melampodium* deserves some explanation. Of all the plants on Earth, why this particular genus? The story begins in 1965 with the start of a Ph.D. study by T. Stuessy, conducted at the University of Texas under the supervision of Prof. Billie Lee Turner. During Stuessy's first week in Austin, Turner led him to the hallway of the fourth floor of the Biological Sciences Building and the many cases of Asteraceae (Compositae), each containing genera housed in folders, with species organized alphabetically. There was quite a bit of *Melampodium*; in fact, the entire right row of shelves contained numerous species of this genus. Turner said: "Now, *Melampodium* is an interesting genus, lots of taxonomic problems, needs careful work. And there is chromosomal variation, too, which should be significant in the systematics and evolution of the group. How would you like to do your thesis on this genus? The group is largely Mexican in distribution, and I know you wanted to do research with plants from that region." Being terribly naïve about research at this point, Stuessy simply said: "OK. I'll do it."

The selection of *Melampodium* for systematic research was not done after months of careful literature searching, evaluating alternative genera, and discussing bold biological questions. No hypotheses were formulated; it was just very simple. From 1965 to 1968, Stuessy immersed himself in the world of *Melampodium*, finally publishing articles on chromosome studies (1970a) and phylogeny (1971), followed by a comprehensive revision of the genus (1972). Further work extended to related genera (Stuessy, 1969, 1970b), the subtribe (Stuessy, 1973a), and eventually the entire tribe (Stuessy, 1977, 1980).

After graduation from Texas, Stuessy joined the faculty at Ohio State University, Columbus, eventually moving on to examine other Mexican genera of Asteraceae (e.g., *Baltimora*, Stuessy, 1973b; *Lagascea*, Stuessy, 1978; *Isocarpha*, Keil and Stuessy, 1981; *Otopappus*, Hartman & Stuessy, 1983). Some collaborative work on *Melampodium* was published with phytochemists, especially Prof. Klaus Fischer and colleagues of the Department of Chemistry of Louisiana State University (sesquiterpene lactones; Seaman et al., 1980) and Prof. Bruce A. Bohm of the Department of Botany at the University of British Columbia (flavonoids; Bohm & Stuessy, 1991,

2001). Some small grants enabled more populational sampling within the white-rayed complex, which many years later was published (Stuessy et al., 2004).

Into the 1970s and 1980s, quantitative methods for assessing relationships became popular, and this led to a cladistic study of *Melampodium* (Stuessy, 1979), based on data accumulated during the revision of the genus. This placed the understanding of phylogeny of the group on a firmer basis. The cladistic investigation was complemented by numerical taxonomic studies (phenetics) in collaboration with Prof. Jorge V. Crisci from the Universidad Nacional de La Plata, Argentina (Stuessy & Crisci, 1984). These quantitative studies demonstrated the robustness of some of the infrageneric units (sections) within the taxonomy of the genus, and they highlighted a few others that were only weakly supported.

Work on *Melampodium* slowed down considerably as Stuessy moved in 1997 to the Department of Botany and Biodiversity Research of the University of Vienna, Austria, where he became involved with research projects of island biology (especially in the Juan Fernández Archipelago, Chile, and Ullung Island, Korea) and speciation and biogeography in the genus *Hypochaeris* of South America and the Mediterranean region. The research environment in Vienna was rich with colleagues pursuing many different botanical avenues from floristics, monography, morphology, DNA, and cytogenetics. The institute had a long tradition in cytology and cytogenetics, and this continues to the present day.

During a departmental seminar in Vienna, it was decided to read the new book, a chapter per week, by Donald Levin (2002): *The Role of Chromosomal Change in Plant Evolution*. As there were more weeks than chapters, additional presentations were solicited, and Stuessy talked about the wide chromosomal variation in *Melampodium*. This generated some interest, and after the presentation, Hanna Weiss-Schneeweiss, a postdoc in the department, said: "This range of chromosome numbers is quite impressive in *Melampodium*, and it might make sense to look at this again with modern molecular approaches, especially involving DNA." After more discussion, it became obvious that modern investigations could indeed reveal evolutionary relationships more precisely, plus indicating pathways (and perhaps also mechanisms) of chromosomal change within the genus.

Thanks to the generosity of the Austrian Science Fund (FWF), including a Hertha Firnberg fellowship to Hanna Weiss-Schneeweiss, substantial three-year funding was obtained that allowed progress on *Melampodium* to be made. This provided money for collecting trips to Mexico and adjacent regions, as well as laboratory studies, including molecular analyses. A good basis for the project was that Stuessy knew the species of *Melampodium*, having collected them back in the 1960s, which meant that it should be possible to collect material (seeds, leaves for DNA work, buds for cytological studies, and herbarium vouchers) without too much difficulty. Two graduate students, Cordula Blöch and Carolin Rebernig, joined the group and each completed their Ph.D. theses within the project. This FWF project was followed by another three-year effort, this time with emphasis on the origin of polyploid species and the biogeography of selected species complexes. A third Ph.D. student, Jamie MacCann, joined the team in this second grant phase to add expertise in phylogenetic analyses. The principals for this work were: Tod Stuessy, coordinator of field studies with our collaborator, José L. Villaseñor, of the National Herbarium (MEXU), Universidad Nacional Autónoma de México (UNAM); Hanna Weiss-Schneeweiss, cytogeneticist; and Gerald M. Schneeweiss as phylogenetic and biogeographic expert. This team produced a series of interesting papers (e.g., Blöch et al., 2009; Rebernig et al., 2010a, b, 2012; Weiss-Schneeweiss et al., 2012; McCann et al., 2016, 2018, 2020).

Despite the many publications on *Melampodium* over the past 50 years, never has this information been synthesized into a whole so that the genus might serve as an instructive case study of plant systematics and chromosomal evolution. It is this interest that has led to the preparation of the present book. It begins by introducing the morphology and habitat preferences of the genus. Species occur all over Mexico, and there are very few regions of the country where one cannot find at least one species. This is not a genus that stays nestled in secluded nearly inaccessible valleys, requiring backpacking for two days to find populations. *Melampodium* does quite well with disturbance and can often survive along fence rows, old agricultural fields, and even around buildings. One of the species, *M. divaricatum* (see Frontispiece), has been successfully transported inadvertently to many other countries, and its vitality and adaptability have led to its being introduced to cultivation in the United States and Europe. Because of the existence of people in Mexico for so many centuries, it makes one wonder what the human impact might have been on the original distributional patterns of species within the genus.

Melampodium, therefore, represents a very good case study of a plant genus in which chromosomal change has played an important role in its evolution, with 13 different chromosome numbers ($n = 9, 10, 11, 12, 14, 18, 20, 23, 24, 27, 28, 30, 33$), one of the longest series among genera of Asteraceae (Watanabe, 2008; Semple & Watanabe, 2009). Furthermore, numerical and structural chromosomal changes have accompanied a radiation of species over the varied Latin American landscape, especially in Mexico, allowing colonization of lowland tropical forests, pine-oak associations, thorn-scrub vegetation, savannas, mesquite-grassland, and arid deserts. The background of morphological and ecological diversity within the genus has been combined with detailed DNA studies of phylogeny and phylogeography, plus cytogenetic analyses.

There is much more that future studies of *Melampodium* can teach us about plant adaptation and speciation. It is a group well-positioned for greenhouse studies to reveal more clearly the isolating mechanisms and genetic affinities of taxa. *Melampodium* has a simple monoecious breeding system, with male (disc) flowers in the center of the flowering head and female (ray) flowers toward the exterior. Removal of the disc during development renders the head functionally female, and the reverse when the developing ray flowers are removed. These simple manipulations, plus adaptability to greenhouse culture and ease of seed germination, greatly facilitate making crosses within and among species of the genus. Some preliminary crossing studies have been accomplished (see Chap. 8), but much more can be done.

So, in large measure, this book is a progress report on understanding evolution within *Melampodium*. Our goal has been to bring together the scattered literature on the genus, plus adding new interpretations, to establish a research platform from which further investigations can yield deeper insights. In the meantime, we hope that the reader finds the genus as instructive and fascinating as we have found it over these many years.

Columbus, OH, USA — Tod F. Stuessy
Vienna, Austria — Hanna Weiss-Schneeweiss
Gerald M. Schneeweiss
Ciudad de México, México — José L. Villaseñor

Acknowledgments

All authors are indebted to many institutions and persons who have helped bring a book into print. This work is no exception. Appreciation must go to the institutions that have provided salaries, offices, herbaria, and laboratories for the conduct of the research and writing of this book: Department of Botany, University of Texas, Austin; Department of Botany, The Ohio State University; Department (now Division) of Systematic and Evolutionary Botany, University of Vienna, Austria; and Department of Botany of the Universidad Nacional Autónoma de México (UNAM). Permissions for leaves of absence from campus for field work were provided for T. Stuessy by the University of Texas, The Ohio State University, and the University of Vienna; and for J. L. Villaseñor, from UNAM. Without this administrative support for research on *Melampodium*, it simply would not have been completed.

Research does not get done without financial resources. We have been fortunate to have received several grants that have made our work possible. The principal source of the initial funding came from the U.S. National Science Foundation, which supported Stuessy for two years during doctoral work on *Melampodium* at Texas (through a traineeship grant to Prof. Frank Blair, No. GB-6914). One field excursion to Mexico was partially supported by an NSF grant to Prof. Howard Arnott (No. GB-1458), and supplemental aid came from Sigma Xi. During postdoctoral studies on Melampodiinae at Harvard University in 1971–1972, Stuessy received support from the Maria Moors Cabot Foundation. At Ohio State, he also received a small institutional grant that allowed more field work (and subsequent cytological studies) on the white-rayed complex of the genus. Major funding materialized in Vienna with two three-year FWF (Austrian Science Fund) grants, the first to Stuessy and H. Weiss-Schneeweiss (FWF Grant No. P18201), and the second to H. Weiss-Schneeweiss, G. M. Schneeweiss, and Stuessy (FWF Grant No. P25131). A small grant was provided from the Commission for Interdisciplinary Ecological Studies (KIÖS) of the Austrian Academy of Sciences (No. P2007-12 to TFS). Additional support was also provided by FWF to H. Weiss-Schneeweiss with a postdoctoral Hertha-Firnberg Fellowship (T-218). Very important is to mention the very strong support from the University of Vienna for the purchase of major laboratory equipment for molecular phylogenetic investigations on *Melampodium* and other projects.

The research grants in Vienna made possible Ph.D. investigations by Cordula Blöch, Jamie McCann, and Carolin Rebernig. They accompanied Stuessy and Villaseñor on field trips to Mexico and the southwestern United States, and they carried out most of the laboratory work on bioinformatics and molecular phylogenetics.

Many other persons over the years also contributed to the success of the research effort, and most are so indicated as co-authors on papers. These include: Katharina E. Bardy, Michael H. Barfuss, Bruce A. Bohm, Jere N. Brunken, Jorge V. Crisci, Nicolas (Klaus) H. Fischer, Sabine Jakob, Tae-Soo Jang, David J. Keil, Jiři Macas, Nicholas J. Matzke, Petr Novák, Renate Obermayer, Arun Pandey, Peter Schönswetter, Fred C. Seaman, Scott D. Sundberg, and Barbara Turner. Without contributions from these excellent colleagues and researchers, our work with *Melampodium* would not have been accomplished.

Other persons and institutions have contributed important help in different capacities. A special thanks goes to Enrique Ortiz of the Department of Botany, UNAM, for assistance with field work, databasing of collected specimens, production of distribution maps, and excellent photographs of species of *Melampodium*, many of which appear in this book. Distributional data come from Stuessy (1972), herbarium records at UNAM, and GBIF. Patrick Alexander, Donovan Bailey, Jorge Calónico, Ismael Calzada, Gabriel Flores, Sylvia Kelso, Michael Lenko, Chris Meacham, Monique Reed, Joaquin Sánchez, John Schoenhofen, Richard Spellenberg, and Hugh Wilson also helped with aspects of facilitating field collections. Technical support in Vienna was also provided by Verena Klejna and Gudrun Kohl. Tom Wendt of the Herbarium at the University of Texas (TEX) kindly permitted us to investigate DNA from collections of the recently described *Melampodium moctezumum* (Turner, 2007) and *M. elottianum* (Turner, 2016). Permits for field collections in Mexico were generously provided for Villaseñor et al. from SEMARNAT and in Costa Rica by CONAGEBIO. We also thank Prof. Hennig Adsersen for providing material of *Lecocarpus* for analysis.

Many of the illustrations in this book have come from our own photographs, figures, and maps. Use of some images has been allowed under: Creative Commons (Figs. 2.4A, 2.9A, 4.11A [Hugh Iltis], 4.11B); Wikimedia Commons (Figs. 2.9B, 4.12A–C); and STM publishers agreement (Figs. 2.2, 2.6, 2.7, 4.9, 4.13, and 8.2; Table 8.2). Several individuals have kindly given permission to reproduce additional images: Paul Fontaine (Fig. 2.4B,C); Reed F. Noss (Fig. 1.1); A. Michael Powell (dedication photo); Ingo Schubert (Fig. 1.3); Richard Spellenberg (Fig. 3.2); and Jun Wen (Fig. 1.2). Marek Brys from WorldAtlas.com kindly allowed use of a map of Mexico (Fig. 4.8).

Permissions have been generously granted to allow the reproduction of figures from the following journals [and organizations]: *American Journal of Botany* [John Wiley] (Figs. 5.3, 5.4, 6.3, 6.11, 7.1, and 7.4); *Arizona Highways* (Fig. 2.1B); *Brittonia* [New York Botanical Garden] (Figs. 5.1, 5.2, 9.5, and 9.6); *Evolution* [Oxford University Press] (Figs. 7.5 and 7.6); *Frontiers in Plant Science* [Frontiers

Media] (Fig. 7.9); *International Journal of Plant Sciences* [University of Chicago Press] (Figs. 5.7, 5.8, 5.9, and 7.2); *Journal of Systematics and Evolution* [John Wiley] (Fig. 1.2); *Madroño* [California Botanical Society] (Figs. 8.6, 8.7, 8.8, and 10.6; Table 8.3); *Molecular Ecology* [John Wiley] (Figs. 9.7, 9.8, and 9.9); *Molecular Phylogenetics and Evolution* [Elsevier] (Figs. 6.4, 6.5, 6.6, 9.10, 9.11, and 9.12); *PLoS One* [Public Library of Science] (Fig. 7.3); *Rhodora* [New England Botanical Society] (Figs. 3.3, 3.4, and 3.5); *Systematic Biology* [Oxford University Press] (Figs. 7.7 and 7.8); *Taxon* [John Wiley] (Figs. 3.1, 6.1, 6.2, 6.7, 6.8, 6.9, 6.10, and 10.7); *Wildflower* [Lady Bird Johnson Wildflower Center] (Fig. 2.1A).

Books do not just appear out of nowhere—they must be produced. For this we enthusiastically thank our editor at Springer Nature, Kenneth Teng, who has been consistently supportive of this book project and made many helpful suggestions for its improvement. We also thank our production editors, Kalai Shahethya, and Mahalakshmi Saravanan, for shepherding our rough files into quality final presentations.

Contents

Chapter 1
Biodiversity and Chromosomal Evolution

The world we call home, our Earth, is a paradise. This blue, white, and green marble hurtling through black space is sheathed by a mantle of oxygen and other gasses that allows millions of life forms to exist. They abound on land, in the sea, and in the air, some being as large as the California big trees, or whales that inhabit the ocean, but others being minute, such as the bacteria and viruses, which are far beyond detection with human eyesight. Somewhere in between are humans, members of our species, that survive by utilizing the energy resources from the sun that have been captured by green plants in the manufacture of carbohydrates. From these building blocks, other organic compounds developed and eventually led to the formation of life forms of increasing complexity that now serve as foods for all animals, including ourselves.

The total of life forms on the planet is its biodiversity. The total number of species is still unknown, but estimates range upward of 8.7 million (Mora et al., 2011) or even more, of which only about 2.1 million have been described formally so far (IUCN, 2022). For proper management of biotic resources for our survival, and for that of other species on the planet, inventorying of these life forms must have a very high priority (Stuessy, 2020a). It is difficult to manage anything effectively when so much data are lacking. The challenge is made even greater because the highest levels of biodiversity reside in countries that are still developing and do not have large amounts of excess money for research activities.

Biodiversity involves species, because these represent basic units of diversity that we can observe and understand most easily, but it also involves their functions and interactions that contribute to the formation of ecosystems. In a broader context, biodiversity can be envisaged as consisting of structural, functional, and compositional dimensions (Fig. 1.1), and these involve different hierarchical levels, such as genes, populations, communities, and landscapes. The compositional aspect is the usual concept of biodiversity, ranging from genetic diversity, through populations, species, and finally ecosystems. But these levels also form more general structures over the landscape, involving genes, populations, habitats, and landscapes.

T. F. Stuessy et al., *Systematics, Ecology, and Chromosomal Evolution of* Melampodium *(Asteraceae)*, https://doi.org/10.1007/978-3-031-81995-7_1

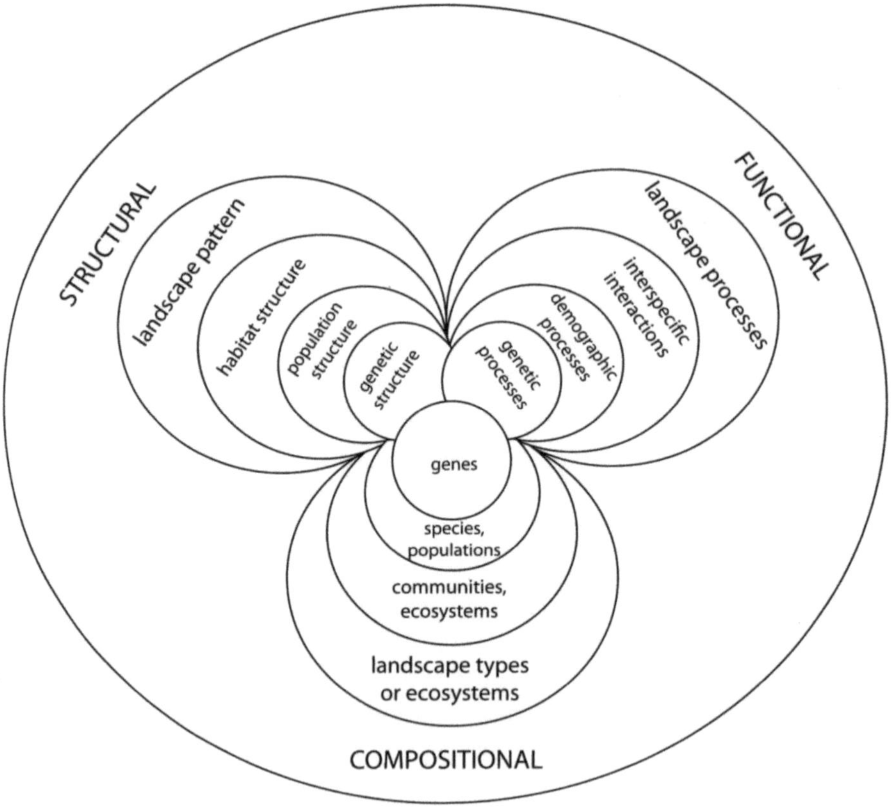

Fig. 1.1 Aspects of biodiversity (structural, compositional, functional) at different levels of organization. (Modified from Noss, 1990: 357)

Furthermore, within these structures are functional aspects that begin with genetic processes and then extend up the hierarchy to include demographic processes, interspecific interactions, and even landscape processes.

The Importance of Biodiversity

It takes little imagination to appreciate how vital biodiversity, in all its dimensions, is for human success on this planet. The basic values of biodiversity deal with goods, information, services, and psychological benefits. As for goods, the amount of harvesting of timber for wood resources worldwide is immense. We remind ourselves that the rooms in which we live are constructed from water-conducting tissues of trees, their secondary xylem. These same tissues are used as fuel in many countries, providing warmth during cold seasons and for cooking year-round. Plants also are used for foods, not only harvested directly from the forests, but also grown in massive factory farms on gigantic scales in the production of wheat, corn (maize), rice, soybean, and numerous others. In addition to foods, plants are extremely important for their fibers that get woven into clothing, plus cords and ropes of a wide variety. Plants also provide medicinals to help alleviate human suffering (Bernardini et al., 2018), with as much as 25% of present pharmaceuticals having been originally derived from natural plant populations (Miller, 1996).

Biodiversity also provides information regarding our natural world. Curiosity is one reason people seek to learn about other species that share our planet, but it is also a desire to manage the environment and increase our chances of survival. The more we can control our environment, the more stable will be our existence, and presumably the happier we will be both now and for future generations. In dealing with microbes, many of which are pathogenic, we need information about their lifestyles, their metabolism, and their reproduction to be able to restrict their harmful effects on us and all other species in the ecosystem.

The services offered by biodiversity come in diverse forms. We are familiar with recreational activities associated with turf grasses in golf courses, football stadiums, and parks, plus the richer organismal diversity presented in zoos and botanical gardens. But even more important is how the individual species interact to create working ecosystems that give us rain, nitrogen fixation, pollination, and carbon cycles. Many studies have been done that suggest that high levels of biodiversity provide greater stability for ecosystem functions (e.g., Tilman et al., 2006; Naeem et al., 2012; Reich et al., 2012; Wang & Loreau, 2016). The yearly economic value of these services is unbelievably high, having been estimated by Costanza et al. (1997) to be somewhere between 16 and 54 trillion dollars (US).

Finally, no one can deny the importance of biodiversity, especially plant diversity, for human well-being. Beautiful flowers are used to celebrate nearly all stages of human life from birth, birthdays, marriage, having children, illness, and finally death. Humans also need contact with animals, as easily witnessed by a large number of domesticated pets, especially dogs and cats, but extending to all other manner of organisms, such as tropical fish, birds, lizards, hamsters, guinea pigs, etc. Humans need this contact with other life forms, and getting out into Nature for an afternoon hike on a path through the woods is extremely therapeutic for shedding stresses at work and home, appropriately referred to as "forest bathing" (Li, 2010; Lee et al., 2011; Farkic et al., 2021). Our species evolved in Nature, and we still have a need to return to these roots, to relax back into our original environment.

But despite the acknowledged value of biodiversity, many challenges exist for its present and future maintenance. The news is full of reports of declining sizes of populations of large animals, such as pandas, tigers, and elephants, suggesting that in future decades these magnificent creatures will only survive in zoos or small reserves, if they survive at all. But the problem goes much deeper, such as the loss of predatory fish in the oceans (Pauly & Maclean, 2003), or the global decline in pollinators (Potts et al., 2010). Regarding plants, it has been estimated that 21–47% of all plant species are now threatened (Pitman & Jørgensen, 2002; RBG Kew, 2016).

The fundamental cause of the loss of biodiversity is the ever-increasing human "footprint" (Kareiva et al., 2007). As of November 15, 2022, the human population reached eight billion persons for the first time in history. With more people come more roads, shipping lanes, and cultivation of land for more food. Human-associated travel has also resulted in the introduction of invasive species of plants and animals (sometimes referred to as the Homogenocene; Samways, 1999) that compete with native species for ever-dwindling resources (Perrings et al., 2005; Powell et al., 2011). Forests have been cut for wood resources and to provide more land for domesticated plant cultivation. The rate of global tropical forest removal has also continued at a high level of approximately six to eight million hectares each year (Achard et al., 2002; Mitchard, 2018). This not only impacts the loss of biodiversity directly, but it also contributes to the broader problem of global warming, which in turn has a greater negative impact on flowering and fruiting time of plants (e.g., Büntgen et al., 2022), which can have consequences for migrating birds that use fruits during their long-distance travels (Both et al., 2006). As temperatures increase, drought can ensue, which will negatively impact amphibians in their aquatic habitats.

Patterns of Biodiversity

Despite centuries of natural history collecting and study, we still have no clear idea of the exact numbers of species that populate our planet. The good news is that the systematic biology community has come together in a rare demonstration of unity around the priority of a global organismal inventory (Anonymous., 1994; Daly et al., 2012). This means getting out into poorly explored regions of the world and making observations and collections. Today, sampling can be of whole organisms or portions of them for DNA laboratory analysis. DNA inventorying, in fact, is essential for learning about microbes, such as in the deep ocean, or fungi, in the soil, for which little morphological information has been accumulated. Sweeping and analyzing the air has even revealed unknown organisms (Johnson et al., 2022). Although much interest toward a global inventory has been provided, successful implementation requires efforts back to the political level, which scientists are often averse to address (Stuessy, 2020b).

Once a general idea of biodiversity for a group is available, usually after decades or even centuries of adequate collecting and study, a more comprehensive investigation is needed to bring all information together and provide a more insightful synthesis. The main publication from these efforts is the biological monograph (Stuessy & Lack, 2011). The general term, monograph, can mean any type of scholarly publication, usually of a large and comprehensive nature. The biological monograph, along with less comprehensive similar publications, the synopsis and revision (Stuessy, 1993), provides the baseline understanding of biological groups. This is the best possible interpretation of the biodiversity of a selected group of organisms at a particular point in time (Borsch et al., 2015; Grace et al., 2021). Floras also contribute to revealing patterns of biodiversity, but their focus is on showing what species inhabit particular areas of the Earth (often political regions, such as countries, states, or provinces) and not on evolutionary relationships.

The biological monograph brings together all available knowledge about a taxonomic group, often a genus, and it weaves this information together with the goals of providing a classification and constructing a phylogeny. These two goals are interwoven, as most classificatory efforts deal directly or indirectly with time-related aspects of evolution of the group. The results are two patterns of biodiversity, the classification and the phylogeny. Without this foundation, it is virtually impossible to infer the processes that produce diversity. Evolution cannot be examined precisely unless we are investigating closely related populations or species, and the clearest presentation of these patterns is in the monograph.

Plant monographs began in the eighteenth century (Stuessy, 2025) and became more common in the nineteenth century and into the first half of the twentieth century. Doctoral training for graduate students used to focus on selected genera (such as *Melampodium*; see Preface), as working out relationships was viewed as good experience for the young plant systematist. It still is a valuable training ground, but with the proliferation of new DNA data, plus the numerous computer programs developed to analyze these data, many young doctoral candidates opt for more analytical projects, especially those involving bioinformatics. This relates to the changing dimensions of plant systematics. Decades ago, an important talent for success was a strong visual memory, which allowed plant images to be stored and intuitively compared with each other, hence facilitating the production of a classification. This ability is still important, but so now is analytical and mathematical ability, favoring students with a strong quantitative background. Furthermore, new Artificial Intelligence (AI) tools are now available to assist with making visual comparisons (Wäldchen et al., 2018; Yigit et al., 2019; Picek et al., 2022).

The most fundamental part of the monograph is the clarification of the limits of species. Many aspects of biodiversity deal with the specific level (especially with governmental regulations), and this level of the taxonomic hierarchy is also fundamental for understanding recognition of genera, families, and higher-level taxa. To investigate species requires adopting a species concept that is workable for the group in question, and this will vary depending on the organisms. The resolution of species limits requires using different lines of evidence (Fig. 1.2), but morphology is always fundamental. The phenotype is what the organism uses for adaptation and interactions with the changing environment, and it is also the level that humans normally use in dealing with species. To these baseline data are added distributions and ecology, to provide the spatial dimension. Some species inhabit very particular habitats, and this is most important to document, especially for conservation purposes. In addition to morphology, distributions, and ecology, other data are brought to bear on understanding the limits of species, such as anatomy, embryology, phytochemistry, DNA sequences, cytology, and cytogenetics. Also important in the monograph is to have the nomenclature resolved, so that consistent and stable names will be available for all species. Biodiversity in a group without an accompanying stable nomenclature is virtually useless for human communication. Armed with consistent species concepts and names, identification keys can be produced that

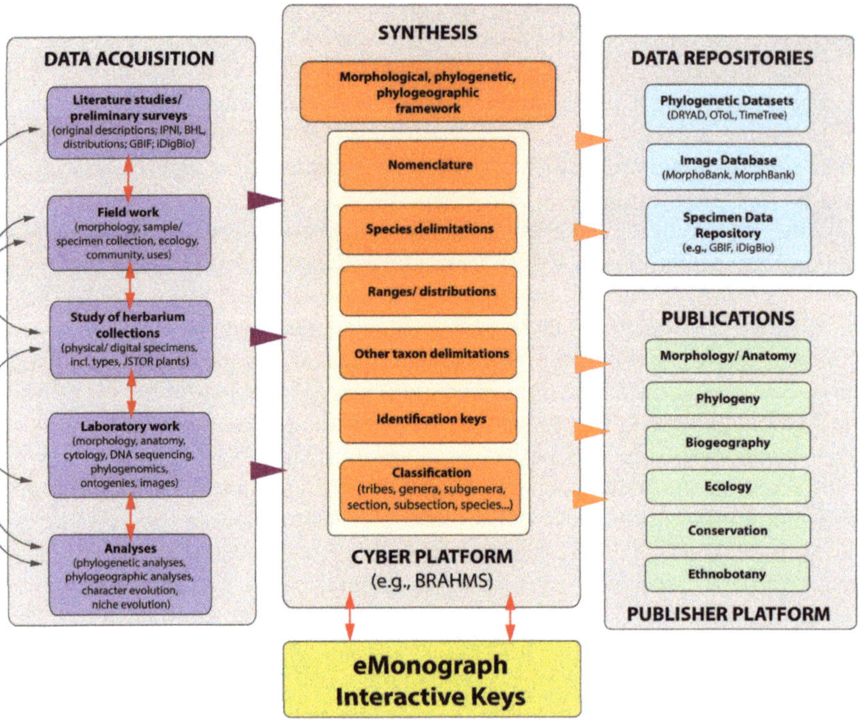

Fig. 1.2 Tasks and workflow in modern botanical monography. (From Wen et al., 2015)

allow other persons to identify specimens in their local areas and for museum workers to identify specimens from faraway regions.

Because of the significance of monographs for understanding biodiversity, and with the realization of the many pressures that biodiversity now faces, concerns have arisen on how monographs might continue to be completed and disseminated more widely. A recent conference was held to address this, and other issues, regarding botanical monographs (Marhold & Stuessy, 2013). A paradox was indicated, wherein young investigators are turning more to DNA analysis and away from monography, which means that fewer new hypotheses of relationships will be developed in the future. The use of Internet resources is one obvious way that more information can be disseminated to the general public at very little cost. Interaction with citizen science groups, such as iNaturalist, should also be encouraged. An Internet monograph can have many photographs of features of the organisms, their habitats, and data on all sorts of other comparative information, including phylogenetic trees. Certain aspects of the monographs can also be designed for the general reader, an encapsulation of relationships and nomenclature, leaving details in other folders for the professional systematist. Collaboration of workers from different subfields on the same genus should also be encouraged, but this involves a psychological shift from one person being the absolute authority in the genus to a consortium of professionals.

Most important is that monography must be creative (Stuessy, 1993). This means that the monographer must be responsible for presenting evolutionary and biogeographic hypotheses as a stimulus for other workers to investigate these aspects more deeply. *Melampodium* is a good case in point. Hypotheses presented in the first monograph of the genus (Stuessy, 1968) have stimulated much further work on evolution and biogeography in the group, as this book clearly documents. A descriptive monograph that presents species with descriptions and distributions is an excellent effort to understand biodiversity, but even better is for the world expert, the monographer, to also think deeper about possible evolutionary tendencies or mechanisms and/or biogeographic ideas and offer these to the reader as a stimulus for more intensive investigations.

Evolutionary Processes in the Origin of Biodiversity

In addition to the importance of understanding *patterns* of biodiversity, equally significant are investigations of the *processes* that have led to the production of new life forms. Through time, populations diverge, and at any one point in time, parts of biodiversity will be in different phases of this differentiation. Some will be just starting their divergence, others nearing speciation or already having completed it, and still others will be trending toward senescence, perhaps even extinction. Learning about the processes that produce biodiversity ties to human curiosity regarding the origin of life on Earth. How biodiversity has already evolved can suggest mechanisms for future modification of domestic plants and animals for human welfare, foods, shelter, clothing, and medicinals. Modern gene manipulation methods will doubtless be very helpful for modifying existing species, but working with natural mechanisms can also be important, and perhaps more acceptable to the general public.

Numerous specific evolutionary mechanisms among different organismal groups have taken place in the production of biodiversity, but two of them are basic: isolating mechanisms, and modes of speciation. Many different types of isolating mechanisms exist, in part due to the diversity of organisms. These are what keep species distinct; the landscape is not just one blurred and mixed gene pool, but instead consists of discrete and mostly stable populational systems. Geographical isolation is fundamental, followed by (and often in conjunction with) ecological isolation (Nosil, 2012). As time ensues after initial speciation, genetic divergence continues until even with secondary sympatry, the population systems maintain their isolation. Recent studies have shown that such divergence can take place even in the face of some genetic exchange between diverging taxa (e.g., Nosil, 2008; Smadja & Butlin, 2011).

Isolating mechanisms in plants are often divided into prezygotic and postzygotic types (Grant, 1981; Levin, 2000; Crawford, 2014; see Chap. 8). The former deal with factors that impede successful fertilization between individuals, such as geography and ecology already mentioned, but also mechanical aspects relating to different morphologies and the development of flowers. A shift in pollinators can bring about reproductive isolation, as can also alteration of flowering time (seasonal or diurnal). If these isolating mechanisms are not efficacious and a hybrid zygote results, then postzygotic factors come into play, being genetic incompatibilities that interfere with development and/or fertility of the hybrid progeny (e.g., Scopece et al., 2008). However, ancient hybridization events in early diverging lineages are known to occur during the speciation process, as documented clearly in the genus *Helianthus* (Owens et al., 2023).

When reproductive isolation results in populations diverging genetically, speciation occurs. The variety of speciation mechanisms is broad because these function within different kinds of organisms. Numerous modes of plant speciation exist (Grant, 1981; Rieseberg et al., 2003; Rieseberg & Willis, 2007; Bock et al., 2023), many deriving from geographic and/or ecological isolation, where populations become separated and diverge from each other. This is allopatric speciation, and it is one of the most common modes in plants and higher animals, especially in the latter. Some workers (e.g., Mayr, 1954; Lewis, 1966) have referred to a process called peripatric speciation, whereby peripheral isolates diverge from the parental gene pool, essentially also based on geographic isolation but on a localized scale. Parapatric speciation is a newer concept (e.g., Coyne & Orr, 2004) whereby two populations can diverge genetically even though some gene flow is maintained between them (e.g., Blanckaert et al., 2020). Sympatric speciation, where a new species originates within the geographic range of the other, is conceptually challenging (Kawecki, 2004) and much less common than allopatric modes, but there are instances of this in plants. One example occurs in the orchid genus *Ophrys* (Schlüter et al., 2009), which has very specialized pollinators attracted by a floral bouquet that mimics the pheromone of the female. Males approach the flower seeking to mate, and in the process they effect pollination. Mutations can occur in the biosynthetic pathways in the production of these attractant compounds that immediately attract another species of insect, hence establishing reproductive isolation (Sedeek et al., 2014). Another mode of speciation that can lead to sympatric origin of new species is via polyploidy, or chromosome set(s) multiplication, so frequent in *Melampodium* (to be discussed below).

Chromosomal Change and Plant Biodiversity

New plant species are produced through many different evolutionary processes, but one that contributes significantly is chromosomal change. Plants can easily tolerate whole-set chromosomal multiplications, as well as dysploid chromosome number changes (i.e., change in chromosome number, mostly due to fusion or fission, without change in gene content; Fig. 1.3). Chromosomes are functional units that ensure transmission of hereditary material and are therefore fundamental for maintaining

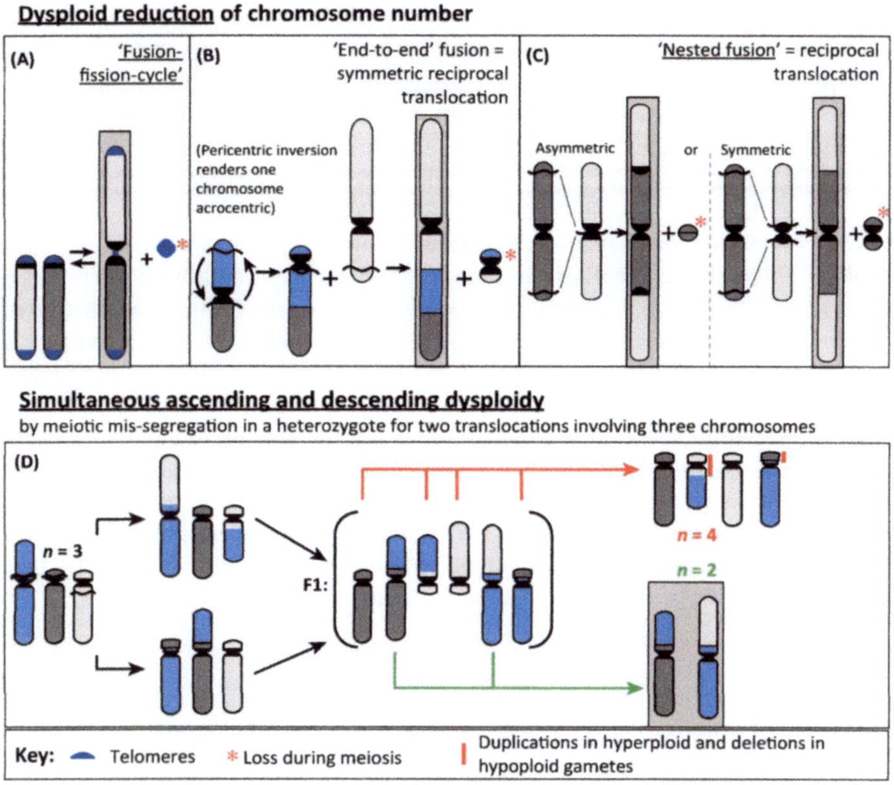

Fig. 1.3 Cytogenetic mechanisms for chromosomal change by dysploid reduction (**A–D**) and simultaneous ascending and descending dysploidy. (**A**) Fusion-fission cycle, (**B**) end-to-end fusion, (**C**) nested fusion, (**D**) translocations and meiotic mis-segregation. (From Schubert & Vu, 2016: 752)

the integrity of species. Modifications in chromosome structure and number play a large role in plant speciation.

All sorts of chromosomal changes can be associated with plant speciation. Some are dramatic and easily observed because they involve whole genome duplications (polyploidy), or change in base chromosome numbers (dysploidy). These types of karyotypic changes have been known for many decades, especially in plants (Stebbins, 1950, 1971). Loss and gain of chromosomes during evolution, involving fusion and fission as mechanisms, have also been investigated in many groups, both plant (Jones, 1978; Levin, 2002) and animal (e.g., King, 1993; de Vos et al., 2020). Other mechanisms are more subtle, but very important, e.g., inversions or translocations (Rieseberg, 2001; Schubert, 2007; Schubert & Vu, 2016). With the advent of genomics, it now becomes possible to connect genetic sequences with chromosome rearrangements to obtain a much more informative view of what has happened during speciation (e.g., Potter, 2017). The tools are now in hand to understand more precisely the genetic bases and implications of chromosomal change, both involving macro- and micro-structural modifications (e.g., Deanna et al., 2022).

It is this background that led to investigations on chromosomal change during evolution within the genus *Melampodium* (Asteraceae). Over the past 50 years, studies have been conducted on the genus, and chromosomal data have always been center-stage. The reason is simply the wide assortment of different haploid numbers: $n = 9, 10, 11, 12, 14, 18, 20, 23, 24, 27, 28, 30, 33$ (base numbers $x = 9, 10, 11, 12, 14$). The following chapters of this book summarize previous results from morphology, distributions, ecology, and karyotypes. To these data have been added molecular phylogenetic investigations that have clarified the evolutionary history of the genus, and this has also made possible inferences of the ancestral chromosome base number. Population genetic studies on the white-rayed complex have revealed the dynamics of evolutionary change involving whole genome duplication (autopolyploidy) within and among populations, as well as serving to interpret biogeographic relationships among the three species of the complex. Analysis of karyotypes within and among species of the genus has revealed new data for all species and a new $x = 14$ lineage. The book shows clearly the importance and potential of chromosomal data in understanding classification and evolution within flowering plants, and *Melampodium* provides a compelling example.

Chapter 2
Melampodium and Its Significance

Step out of the car in Mexico, and you are likely to see a species of *Melampodium*. Most of the 40 species of this group grow comfortably in disturbed areas, such as along fence rows and roadsides, which makes it nearly impossible to avoid the genus. Flowering heads of *Melampodium* are usually yellow or orange-yellow (with three species being white), but this is hardly distinctive with so many other yellow-headed members of the family Asteraceae (Compositae). Especially in August in Mexico, hillsides may be covered by plants with yellow heads, and hiding within this assemblage will likely be one or more species of *Melampodium*. One of the white-headed species, *M. leucanthum*, has graced the covers of two popular magazines (Fig. 2.1).

The meaning of the name, *Melampodium*, has been interpreted in two different ways. Several workers (e.g., Gray, 1884; Cockerell, 1905) have suggested that it was derived from Greek words meaning "black foot," perhaps referring to some fruits that become dark at maturity. The real origin of the name was given by Linnaeus who first described the genus in his *Hortus Cliffortianus* (1737a). He explained in the *Critica Botanica* (1737b: 76) that the name came from "*Melampus medicus graecus*" (Melampus, a Greek physician), who was also regarded as a soothsayer (Wikipedia).

Basic Morphology of Flowering Heads in Asteraceae

To explain what *Melampodium* is requires explaining how it looks, or the morphology, which necessitates a brief introduction to the structures and terminology within the family (Fig. 2.2). Among the more than 23,000 species of Asteraceae (Funk et al., 2009), the variation of form is astounding, but there are still generalities of structure that hold the family together. In fact, it is one of the most easily

T. F. Stuessy et al., *Systematics, Ecology, and Chromosomal Evolution of* Melampodium *(Asteraceae)*, https://doi.org/10.1007/978-3-031-81995-7_2

Fig. 2.1 *Melampodium leucanthum* has been featured on the covers of two magazines. (**A**) *Wildflower* (summer, 2015); (**B**) *Arizona Highways* (Jan 1966, along with *Echinocereus engelmannii*). The species did not receive its proper recognition in **B** because it was listed erroneously as *Zinnia pumila*

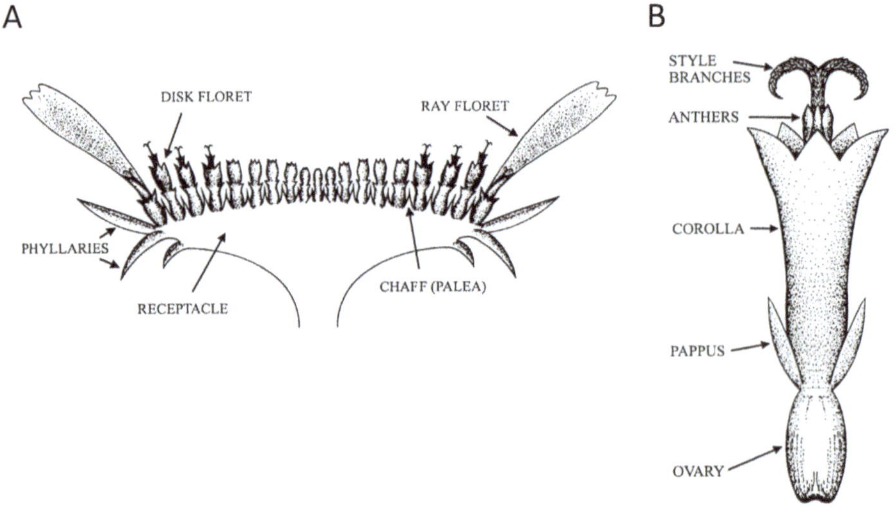

Fig. 2.2 Typical morphology of a radiate flowering head in Asteraceae (**A**) and of an individual regular disk floret (**B**). (From Bohm & Stuessy, 2001: 5, 7)

recognizable families of flowering plants with a long taxonomic history extending back centuries (Stuessy, 2025).

One of the distinguishing features of Asteraceae deals with the inflorescence or flower cluster. Small flowers (florets) are aggregated together (a composite) into a *head* (or *capitulum*), sufficiently tight to give the appearance of a single flower. Heads can also be found in other plant groups (e.g., Calyceraceae, Dipsacaceae, Rubiaceae), but this feature is truly a hallmark of Asteraceae, involving numerous structural and reproductive variations for pollination, defense, and dispersal. Variations in these features no doubt have been largely responsible for the amazing evolutionary success of the family worldwide, occurring on all continents except Antarctica (Funk et al., 2005). Within the family, the number of florets in a head can vary from 1 to more than several hundred.

Surrounding the entire head, to the outside of the strap-shaped (ray) florets, are involucral bracts (or phyllaries; Fig. 2.2A) and these occur in amazing variations throughout the family. Sometimes they occur in numerous overlapping series, perhaps hard or scarious, at other times few and herbaceous. Phyllaries can be colored, covered with pubescence, fused or free, and with entire or dissected margins.

Heads can have flowers that are all the same in morphology or have strong differentiation (Fig. 2.2A). In some members of the family, there is no structural differentiation among the florets within the head, yielding *ligulate heads* (as in the common dandelion, *Taraxacum*, with strap-shaped corollas), or *discoid heads* (as in boneset, *Eupatorium*, with all regular corollas). Even when the florets within a head are uniform, the morphology of florets can vary considerably from group to group (Funk et al., 2009; Stuessy & Urtubey, 2006), particularly in color, size, length, and division of corolla lobes. Even more importantly, there can be differentiation between the florets within the same head. Those in the center of the head are usually radially symmetrical (and hermaphroditic), called the disc (or disk) florets, and those around the periphery, the strap-shaped ray florets (female or neuter), often in one series but sometimes two or more. A ray floret has the corolla modified into a strap-shaped structure, being the fusion of three of the corolla lobes; these extend outward and ring the central disc florets. Heads that have both ray florets and disc florets are called *radiate heads*. A useful mental image of Asteraceae would be the cultivated sunflower (*Helianthus*). This well-known plant has conspicuous ray florets toward the outside of the head and numerous disc florets inside. In *Helianthus* (and *Melampodium*), thin chartaceous bracts (chaff or paleae) are interspersed among and subtend the ovaries of the disc florets, likely offering protection from predatory insects (Stuessy & Spooner, 1988). Other than presence/absence, variations of structure of paleae usually have relatively low taxonomic value.

The structure of a typical disc floret in the family consists of a fused corolla of five petals (3 or 4 are also known; Gardner, 1977), which contains the reproductive structures inside (Fig. 2.2B). The stamens are five, and each one has a small filament attached to the inside of the corolla wall. The anthers are fused into a tube (hence the often-used term for study of the family: synantherology). Rising from inside the base of the corolla tube is the style, often surrounded by a small ring-like nectary, and with a brush-like stigma that pushes up through the anther column during pollen presentation. The stigma lobes usually open outward after pushing out the pollen, exposing surfaces receptive to pollen. Variations in stigma morphology can be very diagnostic across the entire family and at different levels of the taxonomic hierarchy.

Attached between the corolla and inferior ovary is the pappus, which can occur in an amazing variety of structures within the family, such as awns, spines, feather-like arrays (plumose), and even sometimes being absent (as in *Melampodium*). The pappus may also be deciduous, falling off before the fruits mature. The function of the pappus is dual (Stuessy & Garver, 1996): protection of the ovaries from probing insects, and especially if plumose, dispersal by wind after the fruits mature. A familiar example of the latter is the dandelion, in which the apex of the fruit elongates after it matures, and the pappus flattens out to provide an efficient parasol for dispersal.

Diagnostic Features of *Melampodium*

Of all the 1300 genera of Asteraceae (Funk et al., 2009), it is possible to morphologically diagnose the genus *Melampodium* without too much difficulty. They are mostly herbaceous plants, with a few being subshrubs. The heads are radiate rather than ligulate or discoid (Fig. 2.3A), which eliminates lots of the diversity (tribes and subtribes) within the family. A critical character is having bracts (chaff or paleae) on the receptacle, which narrows the field even more, and if we add that the genus has a New World distribution, we begin to focus on the Heliantheae alliance (Baldwin, 2009).

In *Melampodium* the heads are radiate, and there is a marked structural difference between the two types of florets. Those toward the periphery of the head are female only (lacking anthers), having a single ovary that matures into one seed (as with all Asteraceae), and they have ray corollas, usually conspicuous but sometimes short and inconspicuous. The small tubular florets in the center of the head (the disc florets) are male only; they produce viable pollen but cannot set seed. In these florets, due to sterility of the female parts, the stigma is only brush-like and never

Fig. 2.3 Diagnostic morphological features of *Melampodium*. (**A**) Flowering head of *M. leucanthum* showing eight female fertile ray florets with white corollas and yellow stigma lobes, and inner sterile disc florets (five at anthesis in the photo) with yellow corollas, brown anthers, and protruding undivided style branches. (**B**) Underside of flowering head of *M. montanum* showing five phyllaries forming the typical cup-shaped involucre. (**C**) Single fruit of *M. divaricatum*, revealing sculpturing of the inner phyllary that encloses the fruit

divides. Its only function is to push pollen out through the anther tube to be available for visiting pollinators (insects). The reproductive condition of the head, therefore, is simple monoecy, that is, consisting of both separate male and female flowers on the same plant. This monoecious breeding system is unusual within the family (Crawford et al., 2023).

The leaf-like bracts (or phyllaries) that surround the florets of the head are arranged in two series: An outer herbaceous whorl of 2–5 bracts (Fig. 2.3B); and an inner series equal to the number of ray florets and completely enclosing and fusing with the fruits of each one (Fig. 2.3C). At maturity the inner bracts remain fused to the fruits, providing protection and in some cases, due to awns or hoods, aiding in dispersal. The morphological configurations of these fruit and bract combinations (usually collectively referred to as "fruits") can be taxonomically very diagnostic.

A series of anatomical studies by Prof. K. S. Manilal from the University of Calicut back in the 1960s and 1970s (Tiagi & Manilal, 1964; Manilal, 1973, 1975) has claimed that the flowering heads in *Melampodium* and the related genus *Acanthospermum* (another allied genus, *Lecocarpus*, was not examined) are secondarily headed. In Asteraceae, several genera do have secondary heads, such as *Lagascea* (Stuessy, 1978), and these are easily observable by florets having their own small involucres. Manilal has suggested that the phyllaries enveloping the ray florets in *Melampodium* consist of two fused bracts, which provided him evidence that each ray floret represented a separate single-flowered head. He considered all the disc florets taken together as another separate head. The critical point for Manilal was that one can observe two sets of vascularizations in the enclosing phyllaries (he examined six species of the genus). The difficulty with this interpretation is that the vascular traces of the inner phyllaries could just have easily divided during evolutionary fusion with the achenes. Another point is that in some other genera, such as *Smallanthus* (also in the Heliantheae alliance), the inner involucral bracts partially enclose (but do not fuse with) the achenes, which indicates that such a trend of envelopment also occurs in other members of the family.

In the current comprehensive classification of Asteraceae (Funk et al., 2009), *Melampodium* belongs to the Heliantheae alliance in tribe Millerieae with the closest generic relatives being *Acanthospermum* and *Lecocarpus* (Panero, 2007; Fig. 2.4A–C). These three genera have long been associated together (e.g., Stuessy, 1977), and all of them have the diagnostic features of having male-only disc florets and the inner series of involucral bracts enveloping the ray fruits. In *Melampodium* the mature fruits are smooth, wrinkled (Fig. 2.3C), pitted, and may be crowned with a hood or awn. In *Acanthospermum* (Fig. 2.4A) the fruits are short-spined with or without hooks, and in *Lecocarpus* (Fig. 2.4C) they are dramatically spiny or winged. *Acanthospermum* with six species (Stuessy, 1970b) occurs in Central America (perhaps just getting into southern Mexico; Turner, 2017) and down into Andean South America, and *Lecocarpus* with four species is restricted (endemic) to the Galápagos Islands (Eliasson, 1971; Wiggins & Porter, 1971; Adsersen, 1980; Tye & Jaramillo, 2022).

Fig. 2.4 Close generic relatives of *Melampodium*. (**A**) *Acanthospermum australe*; (**B**, **C**) *Lecocarpus darwinii*, in flower and fruit

Distribution and Ecology

Melampodium not only occurs in Mexico but also the southwestern United States, Central America, and northern Colombia (Fig. 2.5). In Mexico the genus demonstrates a very broad ecological range. A conspicuous aspect has been its ability to radiate into many of the ecological zones of the country, such as tropical evergreen and deciduous forests, arid thorn scrub, pine-oak forests, savannas, mesquite-grasslands, and deserts. The elevational range goes from near sea level (*M. diffusum*, *M. rosei*) to over 3000 m in pine-oak forests (e.g., *M. montanum*, *M. repens*). Collections of *M. paniculatum* have also been reported from Brazil, perhaps introduced, as the species is centered in southern Mexico and Central America. *Melampodium divaricatum* is also known from Brazil, but as this weedy, attractive, and adaptable species is also known from Cuba, Myanmar (Burma), Puerto Rico, and the Virgin Islands (Stuessy, 1972), it is likely to have been introduced in all these regions.

Phytochemical Diversity

Several decades ago, considerable interest was generated in examining the secondary metabolites of plants, in part for possible medicinal import. Numerous compounds have been extracted and characterized from flowering plants, such as

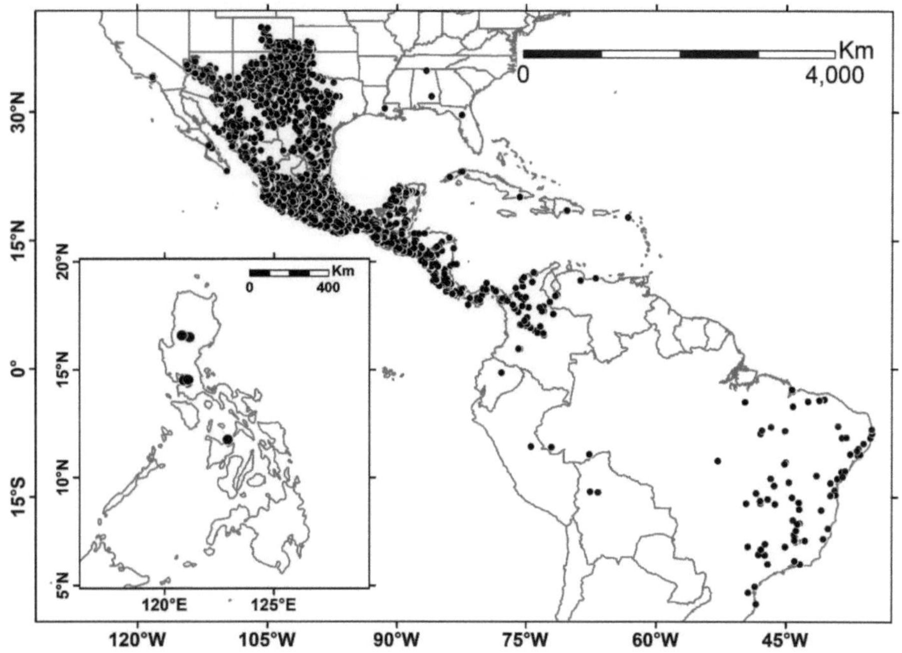

Fig. 2.5 Distribution of *Melampodium*. Inset is of the Philippine Islands (*M. diffusum*). Introduced collections of *M. divaricatum* from Myanmar (Burma) and Puerto Rico not shown

flavonoids, mustard oils, alkaloids, betalains, and sesquiterpene lactones. Flavonoids received much attention from systematists, in part because the methods of extraction and identification were reasonably easy and inexpensive (e.g., Mabry et al., 1970; Markham, 1982), they are under stable genetic control, and they successfully indicated evolutionary relationships, especially in Asteraceae (Bohm & Stuessy, 2001). Flavonoids are phenolic systems (Fig. 2.6) that are found in plant cells, usually in the vacuoles, and serve various functions such as in antibacterial or antiviral activity, UV screens, floral pigmentation, and involvement with basic metabolic processes (Bohm, 1998). Sesquiterpene lactones were also employed (Fig. 2.7), but the laboratory techniques are more complex, requiring a background in organic and secondary products chemistry. These compounds are known to be bitter and repellant to many insects (Burnett et al., 1974), which made them valuable for investigation as possible anti-feeding deterrents in crop plants. General surveys (e.g., Yoshioka et al., 1973; Seaman, 1982) indicated a wide chemical variability (chemodiversity) within Asteraceae.

Melampodium was investigated for flavonoids in some of the species. Stuessy (1971b) analyzed flavonoid spot patterns (from 2-D paper chromatography of plant extracts) in the white-rayed species complex and found a clear distinction between *M. cinereum* and *M. leucanthum*, as well as some infraspecific variation in the latter species. Bohm and Stuessy (1991) investigated five species of *Melampodium* (*M. americanum*, *M. aureum*, *M. bibracteatum*, *M. divaricatum*, and *M.*

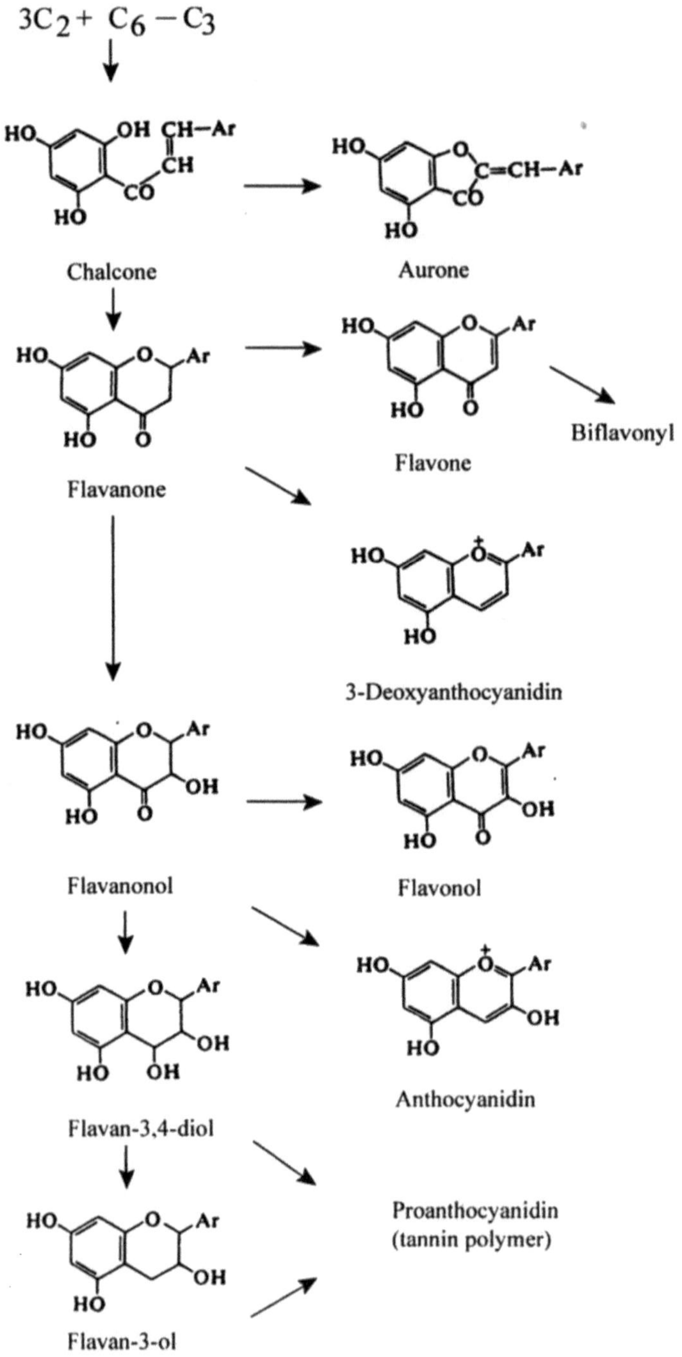

Fig. 2.6 Major classes of flavonoids and their biosynthetic pathways. (From Harborne & Turner, 1984: 134)

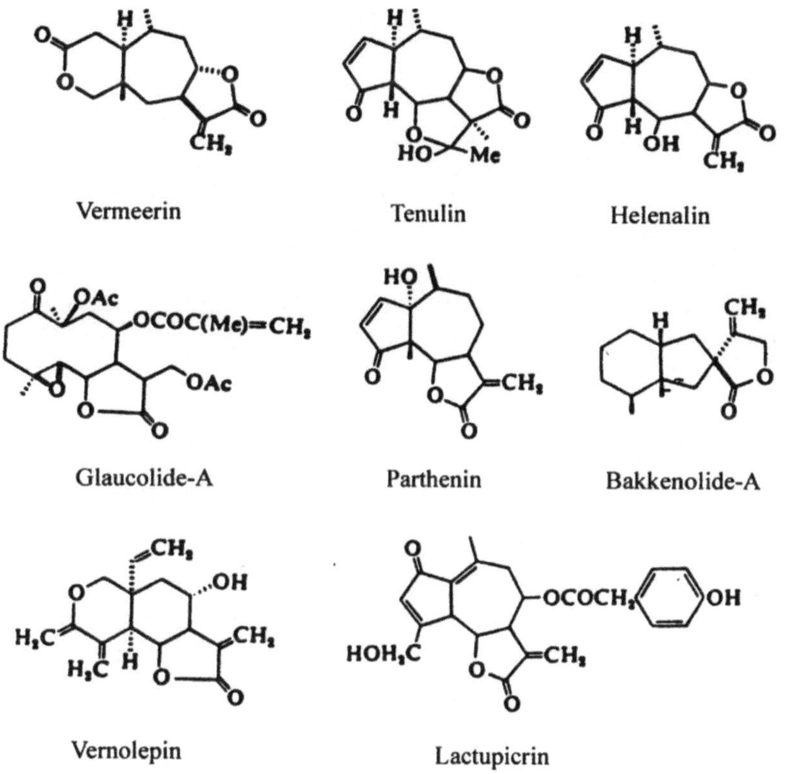

Fig. 2.7 Structures of representative sesquiterpene lactones. (From Harborne & Turner, 1984: 111)

longipilum), representing a good cross-section of the genus. Mostly common flavonoids were encountered, i.e., quercetin, kaempferol, and quercetagetin, but with some differences in substitution patterns on the ring systems. Sampling of 15 individuals within *M. divaricatum* resulted in uniform flavonoid profiles, giving confidence that these compounds are under genetic control and therefore reliable for taxonomic purposes.

Seaman et al. (1980) reported on sesquiterpene lactones in the subtribe Melampodiinae, and this included 10 species of *Melampodium*. Most chemical diversity occurred in Sect. *Melampodium*, and here in two general groups of species. The first consisted of *M. americanum*, *M. diffusum*, *M. pilosum*, and *M. linearilobum* (Sect. *Melampodium* Ser. *Melampodium*), the initial three containing mostly melampolides and the last germacranolides. The second group of species consisted of the white-rayed complex, *M. argophyllum*, *M. cinereum*, *M. leucanthum* (Sect. *Melampodium* Ser. *Leucantha*), and the sesquiterpene lactones are unusual dilactones plus melampodin. Within *M. leucanthum*, populational differentiation exists between the Pecos River region and the rest of the range of the species, with cis-cis-germacradienolides being diagnostic in the former (Seaman, 1982). It is not unusual for species of more arid habitats to have a more elaborated sesquiterpene lactone diversity, as these compounds likely confer defense against predators in these harsh desertic environments.

Broad Range of Chromosome Numbers

Of all the interesting biological aspects of *Melampodium*, one of the most significant is the wide range of chromosome numbers, one of the longest in the family. The following haploid (n) numbers have been reported among 40 species: 9, 10, 11, 12, 14, 18, 20, 23, 24, 27, 28, 30, and 33 (e.g., Fig. 2.8). Also important is that, in general, chromosome number is uniform among populations of a species (only three cases of infraspecific polyploidy and one case of infraspecific dysploidy are known), which allows these data to be relied upon for assessing phylogenetic and taxonomic relationships. In fact, the chromosome number relationships are crucial for understanding evolution within the genus. The numbers alone suggest polyploid relationships (e.g., $n = 10, 20, 30$) that would be fundamental for interpreting speciation as well as phylogeny. A particularly indicative report was in *M. sericeum* (Fig. 2.8D), which has 20 large and 10 smaller bivalents in meiosis. This strongly suggested the possibility of an allopolyploid hybrid origin between species with $n = 20$ and $n = 10$. Another challenge would be to unravel which number might be basic or ancestral

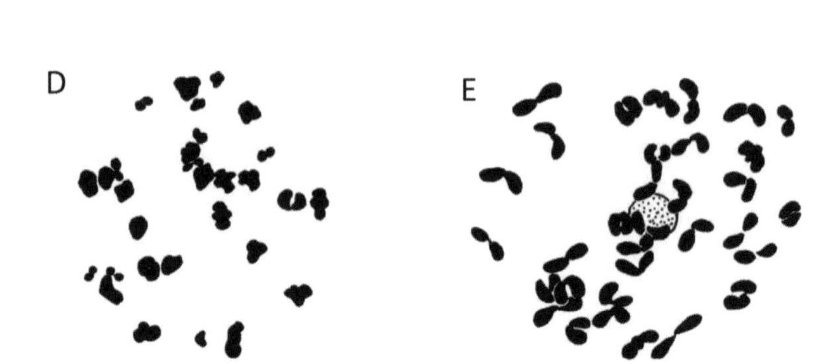

Fig. 2.8 Camera lucida drawings of meiotic chromosome counts of representative species of *Melampodium*. (**A**) *M. americanum* (originally published as "*M. kunthianum*"), $n = 10$; (**B**) *M. montanum*, $n = 11$; (**C**) *M. divaricatum*, $n = 12$; (**D**) *M. sericeum*, $n = 30$ (20 large and 10 small bivalents); (**e**) *M. aureum*, $n = 33$. About ×2000. (From Turner & King, 1962: 266)

for the genus, and how the diploid chromosomal lineages relate to each other through dysploidy. Turner and King (1962) were the first to document chromosome numbers from populations of 26 species of *Melampodium*, and this survey was followed by several others (see Chap. 5 for details).

Horticultural Value

Because of the attractiveness and adaptability of species within the genus, two annuals have been brought successfully into the horticultural trade with very few morphological modifications, i.e., the cultivated forms look like the wild populations. The most broadly distributed species of the genus is *M. divaricatum* (Fig. 2.9A) with a wide range in Mexico in the warmer regions of the country and with extensions into Central America and Colombia of South America. This species has also been introduced, perhaps inadvertently, to other countries of the world, such as Brazil, Myanmar (Burma), and the Virgin Islands (Stuessy, 1972). It grows vigorously, is self-compatible (pers. observ.), and can root easily at the nodes, making it ideal for colonization of new habitats. This species can be found at garden shops under the name *Melampodium paludosum*, which is a later synonym of the earliest valid name *M. divaricatum*.

The second popularly cultivated species of *Melampodium* is *M. montanum* (var. *montanum*; Fig. 2.9B), apparently introduced to the horticultural trade by Hugo Dittmar of Deitingen, Switzerland several decades ago (Uher, 2012). This taxon is

Fig. 2.9 Two species of *Melampodium* that have been successfully brought into cultivation in Europe and the United States. (**A**) *M. divaricatum* (sold usually as *M. paludosum*, a later synonym); (**B**) *M. montanum* var. *montanum* (sold erroneously under the invalid name *Sanvitalia speciosa*)

very restricted in the wild, occurring only in higher-elevation (1740–3360 m; Stuessy, 1972) pine-oak forests in north-central Oaxaca, Mexico. The interesting feature of this taxon is that it was brought into cultivation erroneously as *Sanvitalia speciosa* (or *S. hybrida*), names that have never been validly published (International Plant Names Index). *Sanvitalia* is another yellow-headed member of Asteraceae found in Mexico, and there is indeed a superficial resemblance to *Melampodium montanum*. But the technical features of the ray florets in *Melampodium* with phyllaries enveloping the fruits (achenes or cypselas) serve easily to distinguish the genus from *Sanvitalia* which lacks this feature. Furthermore, the disc florets in the latter genus are fertile, producing seeds. A letter pointing out this misidentification has been sent to the Fleuroselect Association (The Netherlands) that awarded their 2010 Gold Medal to this cultivar *Sanvitalia speciosa* [nom. invalid.](“Million Suns”). They referred the query to the plant breeder (Dittmar), but no reply was ever received. An interesting legal situation might result if another company imports *M. montanum* and begins selling it with the correct name, perhaps even with a catchy trade name of “Mountain Melampodium.”

Outline of Classification

To aid understanding of the diversity of sections and species within *Melampodium* in the subsequent chapters of this book, an outline of the current classification of the genus with chromosomal base numbers (x) is herewith provided (Stuessy et al., 2011). Details of systematics will be covered in Chap. 10, but this outline will help to orient the reader for the chapters to come. The numbering sequence of species follows that from the phylogeny produced with ITS (Blöch et al., 2009; see Chap. 6).

I. Section *Rhizomaria* ($x = 11$)

   1, *M. montanum* (var. *montanum*, var. *viridulum*). 2, *M. aureum*

II. Section *Glabrata* ($x = 11$)

   3, *M. glabrum*

III. Section *Zarabellia* ($x = 9$)

   1. Series *Zarabellia*

      4, *M. longifolium*. 5, *M. mimulifolium*

   2. Series *Tribracteata*

      6, *M. microcephalum*. 7, *M. gracile*. 8, *M. paniculatum*

IV. Section *Alcina* ($x = 11$)

   9, *M. perfoliatum*

V. Section *Bibractiaria* ($x = 14$)

10, *M. bibracteatum.* 11, *M. repens*

VI. Section *Serratura* ($x = 12$)

12, *M. divaricatum.* 13, *M. costaricense.* 14, *M. northingtonii.* 15, *M. tepicense.* 16, *M. sinaloense.* 17, *M. dicoelocarpum*

VII. Section *Nutantia* ($x = 11$)

18, *M. nutans*

VIII. Section *Melampodium* ($x = 10$)

1. Series *Longipila*

    19, *M. longipilum*

2. Series *Cupulata*

    20, *M. cupulatum.* 21, *M. appendiculatum.* 22, *M. moctezumum.* 23, *M. sinuatum.* 24, *M. rosei.* 25, *M. tenellum*

3. Series *Leucantha*

    26, *M. leucanthum.* 27, *M. cinereum* (var. *cinereum,* var. *hirtellum,* var. *ramosissimum*). 28, *M. argophyllum*

4. Series *Glabribracteata*

    29, *M. glabribracteatum*

5. Series *Sericea*

    30, *M. sericeum.* 31, *M. pringlei.* 32, *M. strigosum.* 33, *M. nayaritense.* 34, *M. longicorne*

6. Series *Melampodium*

    35, *M. americanum.* 36, *M. diffusum.* 37, *M. pilosum.* 38, *M. longipes.* 39, *M. linearilobum.* 40, *M. mayfieldii*

Chapter 3
Morphological Diversity

Of all the features that have been used for centuries to delimit plant species, morphology has been paramount. Morphology gives people a convenient handle to remember images of species and use this information in different ways for societal benefit. Every genus of plants containing dozens of species will possess considerable degrees of morphological variation. The person most competent to understand what features are suited for classification within a group is the monographer. Patterns of character distributions across species must be observed, internalized, contrasted, and compared mentally, leading to tentative hypotheses of classification, which are constantly modified as the understanding of the group progresses. In some groups, it is possible to be deceived due to convergent or parallel evolution, and further study will eventually be needed. Fortunately, in *Melampodium*, this has not been problematic.

Despite the obvious utility of morphology for the recognition of species, in this age of DNA data, there are partisans for DNA taxonomy (e.g., Blaxter, 2004; Vitecek et al., 2017), who emphasize using molecular data not only to *detect* new species but also to *describe* them based on differences in nuclear and/or organellar nucleotide sequences. DNA is a powerful tool to indicate relationships because it is here that the evolutionary history of organisms has been preserved, and it is fundamentally useful for groups with cryptic morphology, such as bacteria and fungi. DNA barcoding has also been developed, which uses short genomic sequences for more precise identification of organisms (Hebert et al., 2003; Miller, 2015). But despite these new frontiers, morphology is still needed to provide a convenient visual image of species that can be used in different ways for societal benefit. Furthermore, morphology will always be needed to understand how organisms adapt to their biotic and abiotic environments.

The degree of morphological variation across 40 species of *Melampodium* is remarkable (Fig. 3.1), ranging from annual erect or prostrate herbs, rhizomatous perennials to robust subshrubs, with ray corollas that are yellow, orange-yellow, or white, and heads that vary in size from having fewer than 10 florets to more than

T. F. Stuessy et al., *Systematics, Ecology, and Chromosomal Evolution of* Melampodium *(Asteraceae)*, https://doi.org/10.1007/978-3-031-81995-7_3

Fig. 3.1 Representative species of *Melampodium*. (**A**) *M. aureum*; (**B**) *M. glabrum*; (**C**) *M. longifolium*; (**D**) *M. microcephalum*; (**E**) *M. perfoliatum*; (**F**) *M. bibracteatum*; (**G**) *M. sinaloense*; (**H**) *M. nutans*; (**I**) *M. longipilum*; (**J**) *M. tenellum*; (**K**) *M. leucanthum*; (**L**) *M. glabribracteatum*; (**M**) *M. sericeum*; (**N**) *M. americanum*; (**O**) *M. linearilobum*. (From Stuessy et al., 2011: 437)

110. The challenge of this chapter is to give the reader a sense of the broad range of structural variation that occurs throughout the recognized sections and series of the genus (see outline at end of Chap. 2 and discussion in Chap. 10).

Reproductive Characters

Through years of work with *Melampodium*, it has become obvious that reproductive characters, especially outer phyllaries, ray florets, fruits, and disc florets, contain the most significant morphological taxonomic features. In the discussion below, selected features will be mentioned to show how morphology varies within the genus. Comments will also be offered to indicate which features may be critical for specific recognition. During taxonomic work on the genus, the tendency to formally recognize infraspecific taxa has been resisted. The problem is that there is often so much variation that to achieve a consistent morphological and geographical circumscription of infraspecific taxa would be very difficult. A case in point is *M. americanum*, which displays an amazing array of morphological variations in populations, growing from lowland savannas to mountain pine-oak forests. Admittedly, the ability to quickly diagnose species from reproductive (and other) features within *Melampodium* takes some practice, but with patience, it becomes relatively easy.

External (outer) phyllaries.—The outer phyllaries contain numerous characters of diagnostic value. Their shape is relatively uniform, ovate to narrowly ovate. The number varies from 2 to 5 (occasionally an individual appears in a population with more, presumably a mutation). Most species of the genus have five outer phyllaries; occasionally four are seen, but this is not common or consistent within a taxon. Species with only two outer phyllaries belong to Sect. *Bibractiaria*, which includes *M. bibracteatum* and *M. repens*. These species have very small heads, the latter having only two or three ray and six disc florets. In *M. bibracteatum*, the situation is similar, with (3-) 5-6 ray florets and four disc florets. One might speculate that as the heads diminished in size evolutionarily, perhaps involving a shift in the breeding system from outcrossing to selfing, there occurred a corresponding reduction in number of outer bracts. Three phyllaries are found in three species of Sect. *Zarabellia* (*M. gracile*, *M. microcephalum*, *M. paniculatum*) and in Sect. *Nutantia* (*M. nutans*). The origin of three phyllaries likely occurred in parallel, as *M. nutans* is not closely related to the other three species (Stuessy et al., 2011). Earlier (Stuessy, 1972), this species had been placed in Sect. *Alcina* along with *M. perfoliatum* and *M. glabrum*, based primarily on a similarity of chromosome number ($n = 11$), but cladistic, phenetic, and molecular studies (Stuessy, 1979; Stuessy & Crisci, 1984; Stuessy et al., 2011) have suggested independent sectional status for *M. nutans*.

In addition to the number of phyllaries, their fusion (cohesion) can be useful taxonomically. In the white-rayed complex where the species are quite similar, the best way to distinguish *M. leucanthum* and *M. argophyllum* from *M. cinereum* is that the former two have phyllaries fused ½ to ⅔ their length and the latter only about ¼ their length. This may seem cryptic, but after some experience, this character becomes clear. In species of Sect. *Cupulata*, the involucral bracts are also fused laterally, resembling a small cup.

Another variation that occurs in several species is a scarious apical margin of the outer phyllaries. This can be narrow (but still clearly visible) in Sect. *Cupulata* (*M. appendiculatum*, *M. cupulatum*, *M. glabribracteatum*, *M. rosei*, *M. tenellum*, and *M. sinuatum*), but even more conspicuous in *M. linearilobum* (Sect. *Melampodium*) providing a very important diagnostic marker for this species. It is essential, in fact, for distinguishing it from the related *M. longipes*.

Pubescence (indumentum, vesture) on the outer (abaxial) surface of external phyllaries in the genus varies from glabrous, tomentose, strigose, to pilose. In *M. paniculatum*, glandular hairs are usually present. In *M. divaricatum* the phyllaries are glabrous but ciliate at the margins with short hairs, plus also being flattened (rather than being convex as in other species). Because *M. divaricatum* is extremely variable in leaf features, having these diagnostic characters of the phyllaries is most useful for its identification.

Ray florets.—Ray florets have somewhat fewer characters of diagnostic value within *Melampodium*, but there are some important differences. The most conspicuous variation occurs in the color of the corollas, which in nearly all species is yellow or orange-yellow except for species of the white-rayed complex of the southwestern United States and adjacent Mexico (*M. argophyllum*, *M. cinereum*, *M. leucanthum*). The phylogeny of the genus (see Chap. 6) reveals that the white corollas are a derived condition within the genus, perhaps in some way being adaptive for pollination or solar reflection in the arid regions where these species occur. In some of the smaller-rayed species, such as *M. glabrum*, the rays are a flat yellow, whereas in most other species they are a crisp yellow-orange (e.g., *M. divaricatum*). The rays vary considerably in length, from the very inconspicuous 0.6 to 1.1 mm in *M. strigosum* to as much as 13 mm in *M. aureum.* The number of ray florets also varies from 2–3 in *M. repens* to 11–14 in *M. aureum*.

Fruits.—The fruits in *Melampodium* are actually a complex of a seed, surrounded by a fruit wall (forming the achene or cypsela), and this enclosed completely by an inner involucral bract. Because of the close packing in the head, these fruits are at least somewhat laterally compressed, varying from species to species. Field observations reveal that normally a full or nearly full fruit set occurs. The number of fruits is limited, in contrast to the total number of florets in the head, due to female sterility of the disc florets. At maturity, the fruits are normally brown or dark brown, and completely dry. The number of fruits ranges from 2 to 14 corresponding to the number of ray florets mentioned above, and their size varies from 0.8–1.9 mm in *M. argophyllum* to 4–7 mm in *M. perfoliatum.* The most conspicuous display of fruits is in the latter species, where when fully mature, the disc falls away, leaving a crown of dark fruits (Fig. 3.2).

Fruits also reveal variations in lateral ornamentation. The enclosing involucral bracts can be smooth, ribbed, pitted, or with tubercles. In *M. dicoelocarpum*, as the name might suggest, there are two conspicuous pits on each side of the laterally compressed fruit. The fruits in *M. divaricatum* are very diagnostic, being ribbed in a way that allows certain identification (e.g., Fig. 2.3C). This is especially helpful because of the extensive size and leaf variation of plants in this species, especially after a mowing on the roadside.

In addition to lateral ornamentation, the fruits can bear variable structures at the apex. Fruits are mostly completely enveloped by the inner involucral bracts, but sometimes a "hood" crowns the apex, and in this case, normally a small opening

Fig. 3.2 Crowns of fruits in *Melampodium perfoliatum*. The disc florets have already fallen off the receptacle. The head on the right has mature fruits

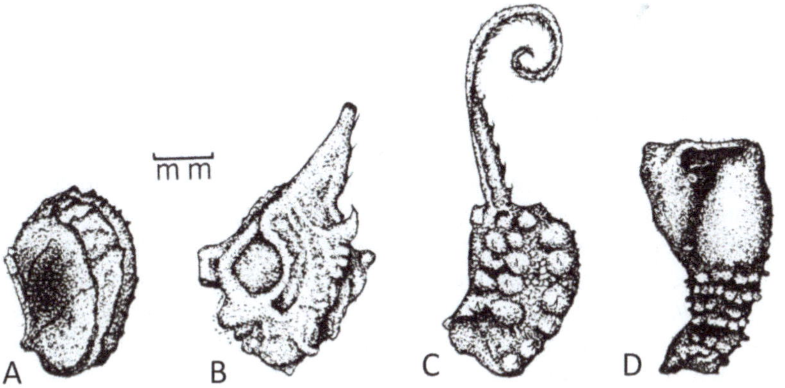

Fig. 3.3 Representative fruits of species of *Melampodium*. (**A**) *M. glabrum*; (**B**) *M. longifolium*; (**C**) *M. longipilum*; (**D**) *M. leucanthum*. (From Stuessy, 1972: 11)

exists around the point of attachment of the ray corolla. These hoods are very characteristic of the white-rayed complex (Fig. 3.3D). In other species, such as *M. longipilum* (Fig. 3.3C) there is an apical awn that attaches toward the adaxial (inner) side of the fruit. Long-coiled awns also exist in *M. longicorne* (as the name implies). Fruit structure can vary within a species, and even within a population (e.g., Fig. 3.4). It is presumed that these extensions of the fruits have some adaptive value for dispersal. Certainly, the long awns do aid in the dislodgement of the fruits from the cup-like involucral bracts. The hooded fruits could aid in wind dispersal, although it might also help harbor moisture in arid regions, perhaps aiding in seed germination. It is worth pointing out that the two related genera, *Acanthospermum* and *Lecocarpus*, also have involucral bracts enclosing the fruits, but these genera possess conspicuous spines, of medium length in the former genus and longer in the latter.

In the wall of the achenes of *Melampodium*, a blackish, hard substance develops. This is phytomelanin, characteristic of many genera of the Heliantheae alliance and

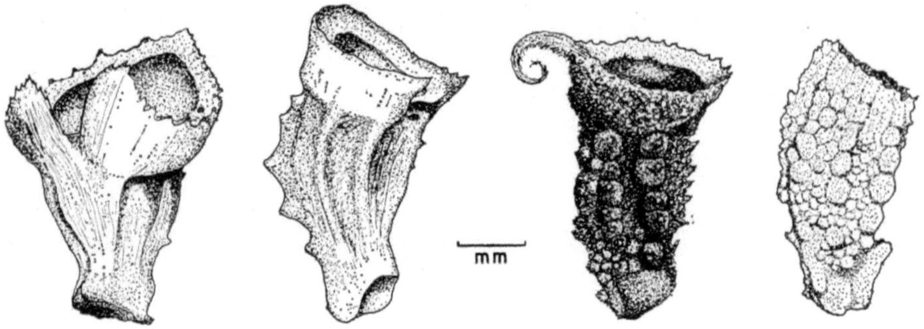

Fig. 3.4 Fruits from a single population of *Melampodium linearilobum* in Michoacán, Mexico (*Stuessy 698*, OS). (From Stuessy, 1972: 11)

Eupatorieae (Pandey & Dhakal, 2001). It is extremely hard and resistant, presumably protecting the embryo inside, and developing in the space between the fibrous layer and hypodermis (Pandey et al., 1989; Stuessy & Liu, 1983). Variations in features of the phytomelanin in mature fruits are mainly the number of striations (thin gaps in the phytomelanin that run longitudinally along the fruit), presence or absence of spines (or pegs), and number of pores (holes in the phytomelanin that develop around the walls of the fibrous layer). Pandey et al. (2015) surveyed genera in the Heliantheae alliance for features of phytomelanin, including nine species of *Melampodium*. Of the taxa investigated (*M. americanum*, *M. bibracteatum*, *M. cinereum*, *M. cupulatum*, *M. divaricatum*, *M. glabrum*, *M. longifolium*, *M. longipes*, and *M. perfoliatum*), only *M. glabrum* and *M. longifolium* lacked striations. In the other species, the number of striations varied between 12 and 25. Pores were common in all species, with 75–125 occurring within a standard measured area of 81 µm. In addition to protecting the embryo inside, phytomelanin with its striations and pores may relate to gas and/or water regulation in and out of the achenes before or after dispersal.

Disc florets.—The florets inside the ring of fruits are functionally male (female sterile). The disc ovaries in *Melampodium* are non-functional but still subtended by paleae. These receptacular bracts generally have a defensive function (Stuessy & Spooner, 1988) in protecting ovaries as well as pollen (in bud) against chewing insects. The disc corollas are regular and consist of five fused petals.

Inside the corolla are five stamens with the anthers fused laterally into a tube. The style grows through this tube and presents the pollen, but no stylar lobes develop in *Melampodium* as there is no receptive surface for pollination and subsequent fertilization. There are two types of abortive ovaries in the disc florets (Fig. 3.5). One type is elongated, being demarcated from the corolla by a distinct annular constriction and present in all species of Sect. *Melampodium*. The other type has no such constriction, and this condition is found in all other sections. One would expect that the more conspicuous abortive ovary might be the ancestral condition and the other derived, but this does not seem to be the case (see Chap. 6). The number of

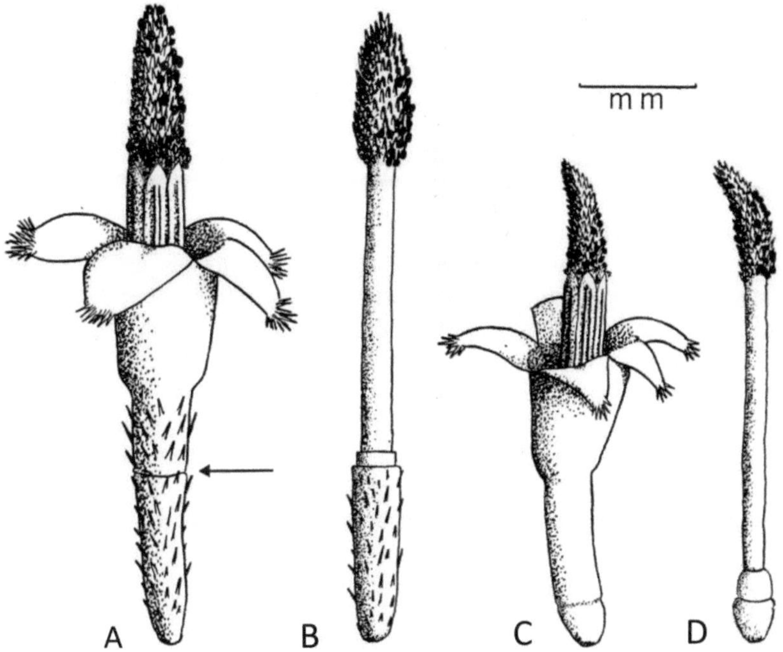

Fig. 3.5 Disc florets of representative species of *Melampodium* showing two different types of abortive ovaries. (**A**) and (**B**) *M. leucanthum* (*Stuessy 752*, OS); (**C**) and (**D**) *M. divaricatum* (*Stuessy 547*, OS). The corollas and stamens in (**B**) and (**D**) have been removed. The arrow in (**A**) points to a marked annular constriction at the point of ovary and corolla tube juncture. Both types of ovaries are capped by ovarian discs (**B** and **D**). (From Stuessy, 1972: 5)

disc florets varies based on the overall size of the heads, ranging from 2–3 in *M. pringlei* to 80–110 in *M. aureum* and *M. montanum* (Stuessy, 1972).

Vegetative Characters

Within *Melampodium*, vegetative features, i.e., leaves, stems, and roots, often provide characters that help distinguish species and serve as indicators of evolutionary relationships. As for habit and duration, most species of the genus are herbaceous annuals, but there are four subshrubs (*M. argophyllum*, *M. cinereum*, *M. leucanthum*, and *M. sinuatum*) and two rhizomatous perennials (*M. aureum* and *M. montanum*). The stem orientation is nearly always upright, except for *M. repens*, which is a ground creeper. Stems can also vary in color (sometimes tinged with purple), adornment (such as pubescence), size (e.g., slender, thick), texture (succulent or not), and even shape in cross-section (e.g., fluted, ridged, winged, etc.). Roots have received less attention historically simply because they tend to be relatively similar and are under a different selection regime below ground. Furthermore, they are often not collected, especially if the species is woody.

Leaves are easy to observe and are nearly always visible on well-prepared herbarium specimens. In *Melampodium*, there is considerable variation among species, and therefore leaf features are very important for distinguishing taxa, in some cases critical, but there can be intraspecific and intra-populational variation. Morphological adaptability and plasticity of some of the species can be surprising, particularly those that have been mowed on roadsides.

The shape of leaf blades can be very diagnostic for species, again, recognizing that considerable variation and plasticity prevails. Leaves in some species are highly divided, such as in *M. linearilobum*, in other species they are lobed (as in *M. gracile*), in two species they have sinuate margins (*M. argophyllum*, *M. sinuatum*), and in many species they are entire. The blade is usually ovate or narrowly ovate, with some species tending toward lanceolate (*M. leucanthum*). The base of the blade is attenuated, except in *M. perfoliatum*, which has distinctive perfoliate leaves.

The indumentum (pubescence) of the leaves can in some cases be very useful taxonomically. In particular, the undersurfaces of leaves in *M. diffusum*, *M. linearilobum*, and *M. sericeum*, and usually also in *M. americanum*, are with long, silky hairs (sericeous). In *M. argophyllum*, the entire leaf surface is covered with dense gray pubescence, but this sort of adaptation is very unusual within the genus. Normally the species have few hairs scattered over the leaf surface, sometimes even completely glabrous (e.g., *M. glabrum*).

Chapter 4
Distribution and Ecology

Species of *Melampodium* are not only very diverse morphologically, but they also occur in many different habitats ranging from lowland deserts and coastal savannas to high-elevation pine-oak forests. The genus is distributed in the southwestern United States (5 species), throughout Mexico (38 species), down into Central America (11 species), and with a small extension into Colombia (3 species). Two species have been collected in southeastern Brazil, *M. paniculatum* and *M. divaricatum*, but it is suspected that these have resulted from introductions perhaps from Guatemala or other parts of the colonial Spanish empire. *M. divaricatum* is the most broadly distributed species of the genus, having been reported as introductions to Burma (Myanmar), Cuba, the Philippine Islands, Puerto Rico, and the Virgin Islands (Stuessy, 1972), plus also isolated appearances in Ecuador and Bolivia. The following discussion provides an overview of patterns of distributions of selected species within the genus; for detailed maps of all species throughout their complete ranges, see Appendix 4 "Habit Photographs and Distribution Maps of All Species (Ordered Alphabetically)".

Geographical Distribution of the Genus

Southwestern United States

Melampodium reaches into the United States in the southwestern states of Arizona, New Mexico, and Texas, plus small extensions northward into the Oklahoma panhandle and adjacent Kansas and Colorado. The most conspicuous group of species is the white-rayed complex, adapted to arid soils and represented in this region by *M. leucanthum* and *M. cinereum*. The former species is widely distributed in limestone-derived soils from Arizona across New Mexico and further east into Texas, halting at the eastern edge of the Edward's Plateau (Fig. 4.1A), plus the

T. F. Stuessy et al., *Systematics, Ecology, and Chromosomal Evolution of* Melampodium *(Asteraceae)*, https://doi.org/10.1007/978-3-031-81995-7_4

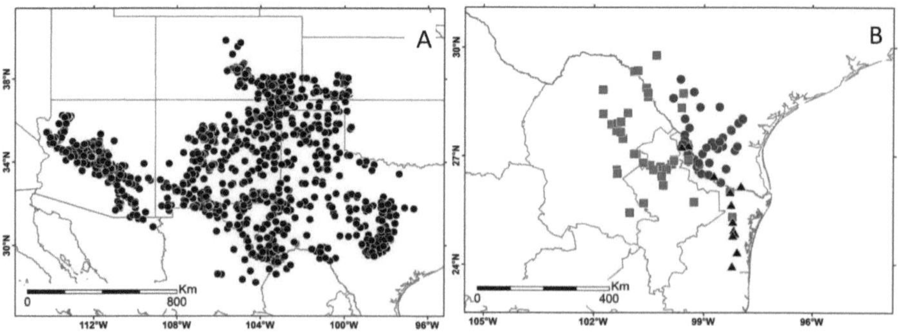

Fig. 4.1 Distributions of white-rayed species of *Melampodium*. (**A**) *M. leucanthum*; (**B**) *M. cinereum* (var. *cinereum*, circles; var. *hirtellum*, squares; var. *ramosissimum*, triangles)

extensions northward already mentioned. The other species of the complex in this region is *M. cinereum* represented primarily by var. *cinereum* in the sandy plains of southern Texas, but also with some populations of var. *hirtellum* and var. *ramosissimum* (Fig. 4.1B), which are more widely distributed in northern Mexico in the states of Coahuila, Nuevo León, and Tamaulipas. The third member of the white-rayed species is *M. argophyllum*, which does not occur in the U.S.A., being restricted to the low arid mountains of southeastern Coahuila and adjacent Nuevo León.

In addition to the white-rayed complex, six other species of *Melampodium* occur in the United States, although only three of them have biogeographical significance. *M. longicorne* is confined mostly to the pine-oak forests of southeastern Arizona where it grows relatively abundantly. It inhabits the higher elevation sky-islands, which are separated by lower, arid, ecological zones. A few collections are also known from New Mexico and across the border in Sonora, Mexico. *M. strigosum* is primarily distributed in Mexico in pine-oak forests scattered along the Sierra Madre Occidental from the state of México northward, but it is also known in Arizona from Cochise and Santa Cruz Cos., then disjunct to Chaffee Co. in Colorado, and Jeff Davis Co. in west Texas. Perhaps the most fascinating distribution is provided by *M. moctezumum*, which was originally described from east-central Sonora, Mexico (Turner, 2007). Only two localities of this species are known, the type locality and another large population from Cochise Co., Arizona. Both sites are in arid zones, in Mexico as thorn scrub and in Arizona as desert scrub. Very significant is that the substrates in both cases are of basaltic lava (Van Devender in Turner, 2007: 259), indicating a very specific edaphic factor that may explain this disjunct distribution. Whether intervening populations might exist between these two localities is unknown. *M. appendiculatum* has been reported from Arizona (Strother, 2006: 36), which would be a short extension across the border northward from its distribution in Sonora, Chihuahua, and northwestern Durango, Mexico. The remaining two species are presumed to have been introduced during historical times. *M. perfoliatum* was collected from around Los Angeles between 1879 and 1902 (Stuessy, 1972: 199), perhaps appearing in ballast deposits from ships coming from Mexico. Since that time, the city has grown tremendously, and in the absence of collections of this species for more than 120 years, it presumably no longer occurs in this region. *M. divaricatum*, the adaptable world traveler into tropical habitats, has also been reported as introduced into Florida (Strother, 2006: 35).

Mexico

Mexico is the center of distribution of *Melampodium* with 38 species being found in all states and provinces except Baja California. Despite the broad coverage of the species across the landscape, the species differ in their distributional patterns, some being very restricted, even extremely local, others having a broader coverage in several states, and still others being more adaptable for surviving in many states of the country. It is important to remember that in Mexico (as in many other countries of the world), people have inhabited the landscape for thousands of years. The Aztec empire in central Mexico was particularly sophisticated with a series of roads that would have encouraged the movement of plants, especially those adaptable to disturbed habitats such as species of *Melampodium.*

Broad distributions.—The most widely distributed species in Mexico are *M. divaricatum* and *M. perfoliatum.* The former is the most widespread species and is found throughout the warmer parts of the country (Fig. 4.2A). The species is largely self-fertile (Stuessy, pers. observ.), and it also can root adventitiously at the nodes, hence making it ideally suited for dispersal and survival. It is no surprise that this species has been introduced into many other countries, and it has also been brought into cultivation successfully (see Chap. 2). *M. perfoliatum* is also widely distributed (Fig. 4.2B), often forming patches on the roadside in tropical habitats. Because it is the tallest species (up to 1.5 m) with large yellow-flowering heads, it is hard to miss. One must be careful, however, because small plants only 20 cm tall have also been reported (Stuessy, 1972). This underlines the importance of fieldwork to see first-hand the infra- and inter-populational variation within a single species.

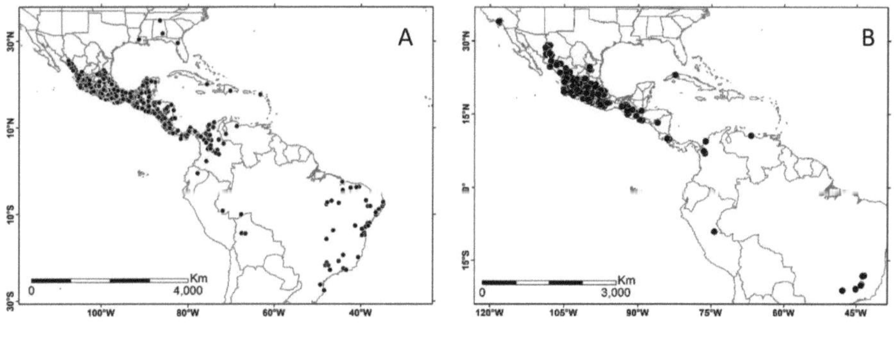

Fig. 4.2 Distributions of the broadly distributed *Melampodium divaricatum* (**A**) and *M. perfoliatum* (**B**)

Intermediate distributions.—Many of the species of *Melampodium* have an intermediate distribution pattern, usually covering several states of Mexico. Within this general category, some species range more widely, viz., *M. linearilobum*, *M. americanum*, *M. sericeum*, and *M. montanum*. *M. linearilobum* runs from Nayarit south into Chiapas (Fig. 4.3A) in both tropical deciduous and pine-oak forests from 20 to 1620 m. Important for this species is a calcareous soil, which often means along road cuts, rocky outcrops, and quarries. The widespread species of the genus tend to have more morphological variation, unsurprisingly, and in *M. linearilobum* the leaves can vary from linear and entire to once or twice pinnatifid. All leaves of whatever shape in this taxon, however, have the characteristic sericeous undersurface and the outer phyllaries have orange-scarious margins. *M. americanum* (Fig. 4.3B) has considerable infraspecific morphological variation. Stuessy (1972: 24) commented on three informal groups: (1) plants with large flowering heads and short leaves with few and broad lobes, in the coastal lowlands near Alvarado, Veracruz; (2) populations with long leaves divided pinnately into long lobes, found in the mountains of Chiapas (and Guatemala); and (3) plants with small heads and a tendency toward the annual habit, in the mountains of the western Sierra Madre in Jalisco, Michoacán, and Guerrero. As might be anticipated from such a diverse assemblage, the elevation range is 4–2175 m, and the habitats vary from savannas to pine-oak forests. *M. sericeum* also has a broad distribution from Sonora in the north to Chiapas and Guatemala in the south (Fig. 4.4A), in numerous pine-oak forests at 1100–2604 m especially along the Sierra Madre Occidental. *M. montanum* (with two varieties) is a rhizomatous perennial found in pine-oak forests at elevations of (610–)1740–3288 m (Fig. 4.4B).

Other species of *Melampodium* also have intermediate distributions but with populations more widely scattered. *M. dicoelocarpum* occurs (Fig. 4.5A) in tropical deciduous forests in southern Nayarit, Jalisco, Michoacán, México, and Guerrero from 158 to 2042 m. The two cytological types (n = 12, 23) do not separate geographically as far as present collections have indicated. *M. diffusum* (Fig. 4.5B) is mainly centered around Acapulco, Guerrero, at low elevations of 30–300 (−1513) m in tropical deciduous forests, but populations have also been located elsewhere

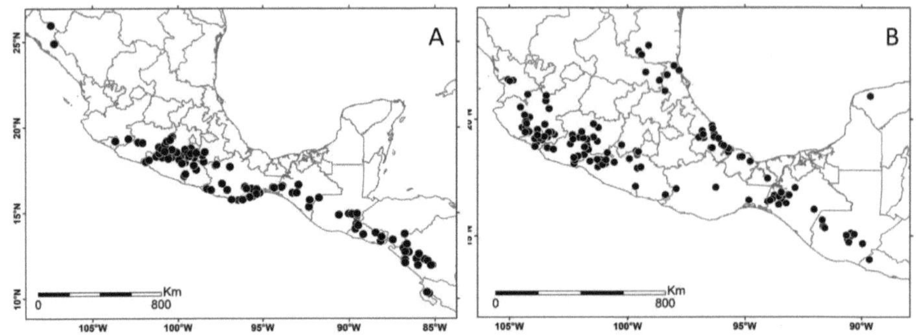

Fig. 4.3 Intermediate distributions of *Melampodium linearilobum* (**A**) and *M. americanum* (**B**)

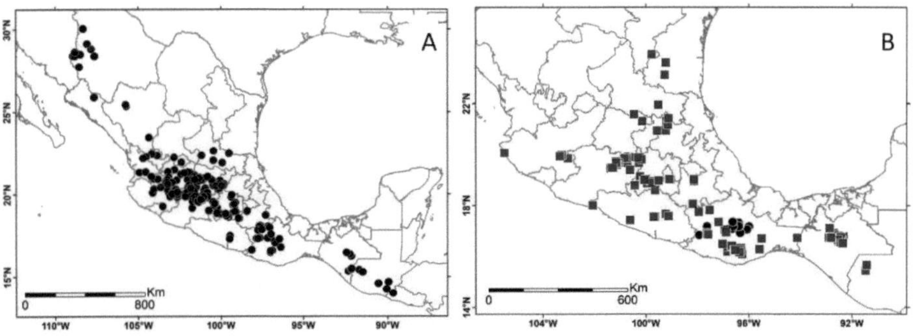

Fig. 4.4 Intermediate distributions of (**A**) *Melampodium sericeum*, and (**B**) *M. montanum* (var. *viridulum*, squares; var. *montanum*, circles)

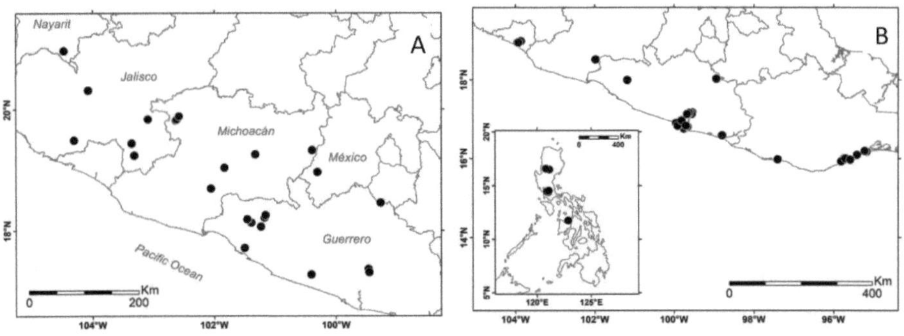

Fig. 4.5 Scattered intermediate distributions of (**A**) *Melampodium dicoelocarpum* and (**B**) *M. diffusum* (inset of Philippine Islands)

along the coast. The species has also been collected in the Philippine Islands, where it probably was introduced during historical times from the active port of Acapulco. Silver from the mines near Taxco was frequently exported to the Philippine colonies (Robinson, 1901; Merrill, 1954).

Restricted distributions.—Based on existing collections, several species of *Melampodium* are very restricted in distribution. *M. mimulifolium* (Fig. 4.6) was previously known only from near the type locality at Totontopec, Oaxaca, collected originally by E. W. Nelson (*740*) in 1894 at 1700–2150 m. More than 110 years later, we encountered a second population of this species (*Stuessy et al. 18656*) c. 130 km south of the type locality. Stuessy (1972) originally registered some doubt that the taxon might be a variant of the relatively common *M. gracile*, but molecular analyses have now confirmed it to be a good species (Blöch et al., 2009). *M. glabribracteatum* (Fig. 4.6B) was described by Stuessy (1970c) from a collection by Conzatti (*5169*) in Ocotlán, Oaxaca, in 1936. Collections in 2013 (*McCann et al. 40*) have revealed that the species is locally relatively common, found in the pine-oak forest zone at 1500–1600 m, and even surviving around a neighborhood school in the city. *Melampodium mayfieldii* (Fig. 4.6) and *M. northingtonii* (Fig. 4.6) are

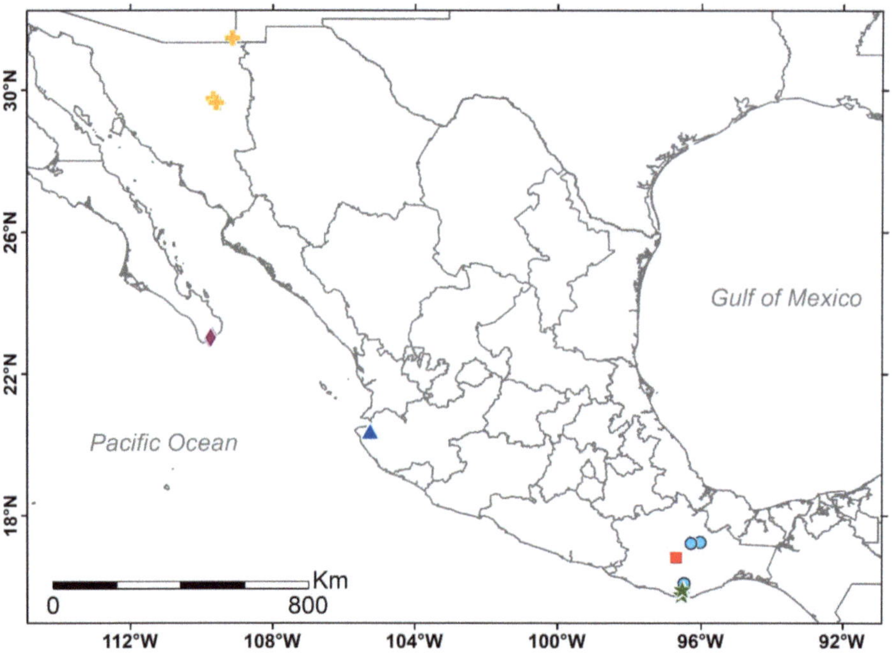

Fig. 4.6 Restricted distributions of species of *Melampodium*. *M. glabribracteatum* (red square); *M. mayfieldii* (blue triangle); *M. mimulifolium* (blue circles); *M. moctezumum* (orange crosses); *M. northingtonii* (green stars); *M. sinuatum* (purple diamond)

two species described recently by Turner (1988, 1993, respectively). Both species are very restricted in distribution, the former in pine-oak forests in western Jalisco and the latter in tropical deciduous forests in southern Oaxaca. *Melampodium sinuatum* (Fig. 4.6) is one of the few perennial species (along with *M. americanum*, *M. aureum*, and *M. montanum*, plus the white-rayed complex). *M. sinuatum* is confined to low (24–70 m) granitic hills at the tip of Baja California, a species now endangered (Villaseñor et al., 2011), but we were able to find a small population (*Stuessy and Villaseñor 19037*) after considerable searching. The final restricted species is *M. moctezumum* (Fig. 4.6), found only in Sonora on volcanic plains with thorn (scrub) forest (Turner, 2007). This same species occurs disjunctly in southeastern Arizona (east of the town of Douglas), also occurring on volcanic basaltic rocky soils.

Central America

Although *Melampodium* is most extensively distributed in Mexico, 11 species occur also in Central America. Two species are concentrated in Central America, *M. costaricense* and *M. paniculatum*. The former occurs in tropical habitats in Costa Rica with extensions north into Nicaragua and south into Panama (Stuessy, 1972; Fig. 4.7A). A few collections are also known from Belize (e.g., *Lundell 6107*) and another one from Colombia (*Pennell 1335*). *M. paniculatum* is found in tropical deciduous forests from 11 to 2257 m in Guatemala, El Salvador, Honduras, and Costa Rica (Fig. 4.7B), with additional recent reports from Nicaragua (Stevens et al., 2001). This species also occurs north into Mexico, especially in the low mountains of Chiapas, plus a transitory introduction into southern California, as well as in Cuba, Colombia, and Brazil.

Four additional species of *Melampodium* occur primarily in Mexico but also extend down into countries of Central America. The widely occurring *M. divaricatum* is unsurprisingly found in tropical habitats in all Central American countries (Fig. 4.2A), extending also down into Colombia and Brazil of South America. *M. linearilobum* (Fig. 4.3A) continues from Mexico into Guatemala, El Salvador, Nicaragua, and Costa Rica, always associated with limestone substrate, and *M. perfoliatum*, well distributed within Mexico, is also occasionally found in Guatemala, El Salvador, Nicaragua, and Costa Rica (Fig. 4.2B). *M. sericeum*, widely distributed throughout the highlands of Mexico, occurs in Guatemala and across the border south into western El Salvador (Fig. 4.4A).

Five additional species dip into Guatemala from their broader distributions in Mexico. *M. americanum* inhabits pine-oak forests, savannas, and tropical deciduous forests throughout Mexico, and is found in the departments of Huehuetenango and Baja Verapaz in Guatemala plus western El Salvador (Fig. 4.3B). *M. longipilum* in Mexico occurs in tropical deciduous and pine-oak forests at elevations of 22–2466 m, being also collected once in Guatemala between Cuilco and Ixmoquí in Dept. Huehuetenango (*Steyermark 50916*). *Melampodium montanum* var. *viridulum* is relatively common in Mexico in pine-oak forests, especially in Chiapas, and this extends down across the border into Dept. Huehuetenango, known there from several collections (Fig. 4.4B). Two closely related species, *M. gracile* and *M. microcephalum* (in Sect. *Zarabellia* Ser. *Tribracteata*), are scattered across Mexico from 30 to 2575 m, mainly in tropical deciduous and tropical evergreen forests. *M. gracile* also extends into the departments of Huehuetenango and Petén of Guatemala as

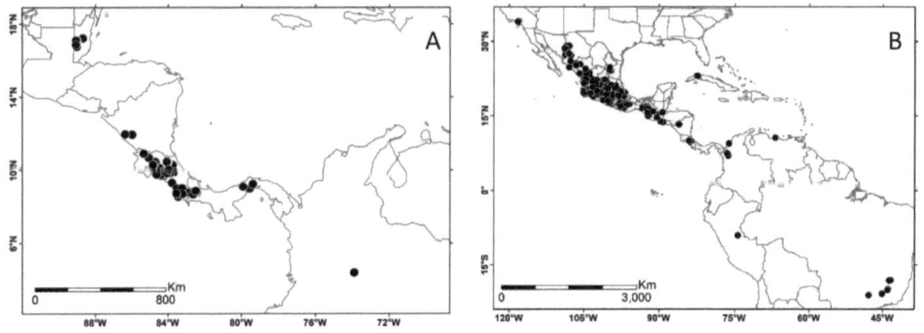

Fig. 4.7 Distributions of *Melampodium costaricense* (**A**) and *M. paniculatum* (**B**)

well as one collection known from Belize (*Lundell 6759*). The close relative, *M. microcephalum*, is known also from Huehuetenango (Stuessy, 1972).

South America

The distributions of species of *Melampodium* in South America are mostly extensions of *M. divaricatum*, *M. paniculatum*, and *M. costaricense* that are known more extensively in Mexico and Central America. *M. divaricatum* is the most widespread of the genus, colonizing many states of Mexico, down throughout Central America in the lower elevation of tropical habitats, and into the northwestern coastal region of Colombia (Fig. 4.2A). *M. paniculatum* is concentrated in its distribution (Fig. 4.7B) in Guatemala, extending southward into El Salvador, Nicaragua, and Panama, and with a few isolated collections (e.g., *Triana 1331*) from Colombia, Ecuador, and Venezuela. It is also found scattered in Brazil in Mato Grosso, Minas Gerais, Rio de Janeiro, São Paulo, and the Federal District. With such a disjunct distribution, it suggests that the populations in Brazil may have resulted from introductions from some ports in Central America during Spanish colonial times. It is interesting, however, that the only known hexaploid chromosome counts ($n = 27$) of this typically tetraploid species ($n = 18$) have come from two populations in Brazil (Coleman, 1970; see also a note in Stuessy, 1972: 172). It is possible, however, that hexaploid populations may also occur in Central America, as very little surveying of this species has been done in this region (so far only analyzed from Guatemala). The last species in South America is *M. costaricense*, known from a single collection (Fig. 4.7A) in Cundinamarca, Colombia (*Pennell 1335*), another possible historical introduction, perhaps from Costa Rica.

Ecological Distribution of the Genus in Mexico

One of the most significant aspects of understanding the evolution of *Melampodium* in Mexico, where the genus is concentrated, is to appreciate the many different ecological zones into which the genus has radiated. Several parameters might be used to express these habitat differences, such as with detailed climatic, physiographic, or geological aspects of the environment, but one frequently used measure is vegetation zones. This is particularly useful, because botanists who have collected the genus have usually described the environment on herbarium labels in terms of associated vegetation. Complex systems of vegetation descriptors exist for Mexico, such as that of Miranda and Hernández (1963), but this is too dissected to help discern patterns that relate to chromosomal lineages within *Melampodium*. The system that is most useful in our context is that from Leopold (1950). Despite its age, it still provides an appropriate ecological background for Mexico in which to infer correlations with the distributions of species and evolution of *Melampodium*.

Leopold first divides the vegetation zones into temperate and tropical, which makes intuitive sense, especially for anyone having traveled the roads of the country. The Tropic of Cancer (23° 26′ N) crosses the country north of Tampico in the East, just above Mazatlán on the western coast, and just above Cabo San Lucas in Baja California Sur (Fig. 4.8). The vegetation can change quickly from one locality to another in Mexico. The country extends from sea level to 5636 m, the highest point being the volcano Pico de Orizaba situated on the border between the states of Veracruz and Puebla. Two massive mountain ranges traverse the country from north to south, the Sierra Madre Occidental along the western side, and the Sierra Madre Oriental in the east, the two joining in the center near Mexico City (Fig. 4.8).

Leopold recognizes five temperate and seven tropical zones (Fig. 4.9; Table 4.1). It is instructive, and perhaps surprising, that the temperate zones together represent 72.2% of the land area of Mexico in contrast to only 27.8% for the tropical regions. The largest zone is the pine-oak forests, comprising 25.8% of the area of the country, followed by desert (23.1%) and mesquite-grassland (21.6%). Boreal forests (0.5%) and chaparral (1.2%) are only minor elements of the landscape. Although covering less area, the topical vegetation zones are diverse and provide many additional ecological opportunities for encouraging speciation. The largest tropical area is covered by tropical deciduous forests with 9.1%. Close behind are the rain forests in the south of the country with 7.3%, followed by tropical evergreen forest at 5.5%,

Fig. 4.8 Generalized relief map of Mexico showing major mountain chains and Tropic of Cancer

Fig. 4.9 Vegetation zones of Mexico. (From Leopold, 1950; redrawn in Pesman, 1962)

Table 4.1 Vegetation zones of Mexico, giving percentage of total land area, approximate square miles, and range of average rainfall

Vegetation zone	% total land area	Area in sq. miles	Range of annual rainfall (in)
Temperate			
Pine-oak forest	25.8	195,600	18–70
Desert	23.1	175,100	2–15
Mesquite-grassland	21.6	163,700	9–36
Chaparral	1.2	9100	14–30
Boreal forest	0.5	3800	30–74
Tropical			
Tropical deciduous forest	9.1	69,000	24–60
Rain forest	7.3	55,300	60–200
Tropical evergreen forest	5.5	41,700	36–112
Thorn forest	3.2	24,300	16–34
Arid tropical scrub	1.3	9800	12–28
Savannah	0.9	6800	34–100
Cloud forest	0.5	3800	70–200

From Leopold (1950: 509)
Total % of temperate zones = 72.2; Total % of tropical zones = 27.8

thorn forest with 3.2%, arid tropical scrub at 1.3%, savannah with 0.9%, and cloud forest with 0.5%. Elevation obviously correlates with vegetation zones. Species adapted to pine-oak forests grow normally from 1000 to 3000 m, and those in coastal savannas from 0 to 300 m.

The different vegetation zones in Mexico have developed in different climatic regimes. Fig. 4.10 compares annual temperature and precipitation for tropical evergreen forest in Veracruz (A), pine-oak forest in Tlaxcala (B), and desert in Chihuahua (C). The tropical evergreen forest is characterized by a relatively constant temperature throughout the year (c. 20 °C) and with high levels of precipitation (c. 400 mm), especially in the summer months. The higher elevation pine-oak forest has lower consistent annual temperature (11–15 °C) and with rainfall reaching over 100 mm from June–August. In deserts, the temperature varies dramatically over the year from about 8 °C in January and reaching to 27 °C in summer. Rainfall is the scarce commodity at c. 10 mm per month except for nearly 40 mm in July.

Species of *Melampodium* are found in seven of these 12 vegetation zones (Figs. 4.11 and 4.12). The high elevation boreal forests and cloud forests are not conducive for habitation by the genus, and no species occur in chaparral in the temperate zone nor in the arid tropical scrub or rain forest of the tropical zones. Coordinate with the pine-oak forests covering one-fourth of the country, 24 species of the genus are found principally in this zone (Fig. 4.13). The next most inhabited vegetation type by *Melampodium* is the tropical deciduous forest with 16 species (at least in part). Four species occur in deserts, three in tropical evergreen forests, three in savannas, two in thorn forests, and one in mesquite-grassland.

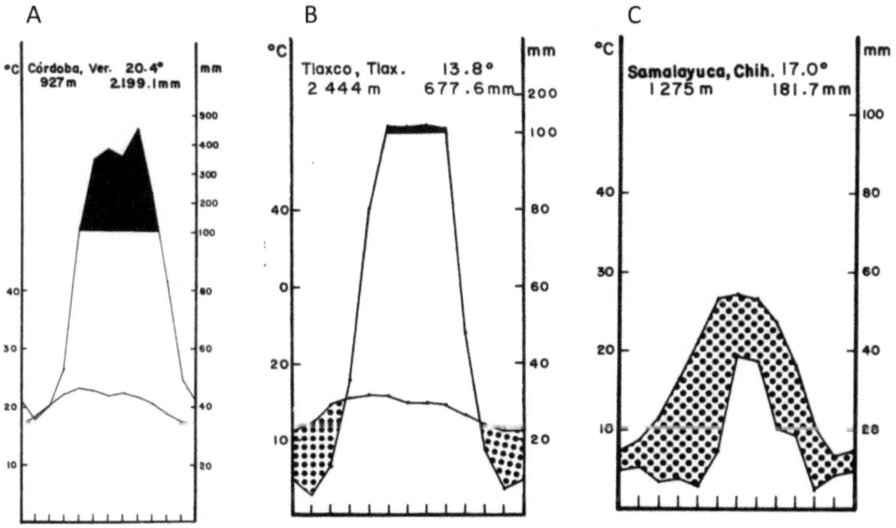

Fig. 4.10 Climate diagrams giving ranges of annual temperatures and precipitation in three vegetation zones of Mexico. (**A**) Tropical evergreen forest (Córdoba, Veracruz); (**B**) pine-oak forest (Tlaxco, Tlaxcala); (**C**) desert (Samalayuca, Chihuahua). (From Rzedowski, 1978: 161, 312, 239)

Fig. 4.11 Tropical habitats in Mexico in which species of *Melampodium* occur. (**A**) Thorn forest (Morelos); (**B**) tropical deciduous forest (Sonora); (**C**) savanna (Nayarit); (**D**) tropical evergreen forest (Nayarit)

Fig. 4.12. Temperate habitats in Mexico in which species of *Melampodium* are found. (**A**) Mesquite grassland (Sonora); (**B**) pine-oak forest (Chihuahua); (**C**) desert (Sonora)

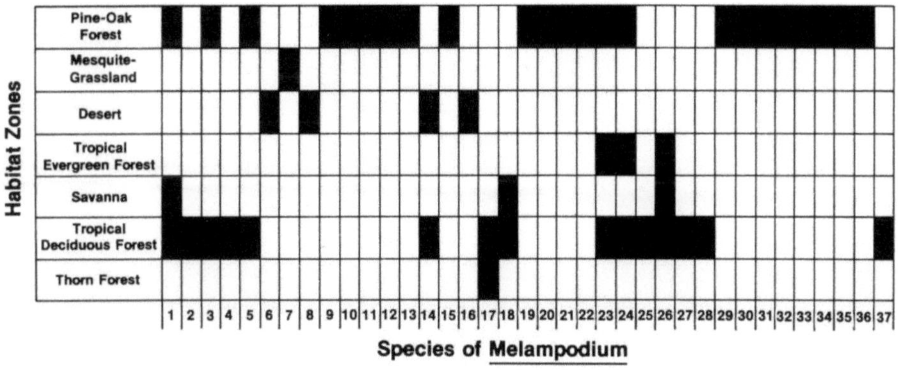

Fig. 4.13 Occurrence of species of *Melampodium* in vegetation zones of Mexico (after Leopold, 1950). Numbers assigned to species are from the classification of Stuessy (1972). New species described recently by Turner (1988, 1993, 2007): *M. mayfieldii*, pine-oak forest; *M. moctezumum*, thorn forest (scrub); *M. northingtonii*, tropical deciduous forest. (From Sundberg & Stuessy, 1990: 85)

Individual species vary in their adaptability to different ecological zones (Fig. 4.13). Four species have been able to establish in three different ecological zones: *M. divaricatum* (unsurprising, due to its weedy nature), *M. americanum*, *M. gracile*, and *M. microcephalum*, and all of them are quite widely distributed within Mexico. Five species occur in two vegetation zones (*M. cupulatum*, *M. linearilobum*, *M. pilosum*, *M. rosei*, and *M. tenellum*), and 31 species are confined to only one vegetation type. The clear message is that most species of the genus have been reasonably successful in colonizing regions of the country, but most within the same ecological zone. The principal home of the genus in context of specific diversification would seem to have been within the pine-oak forest zone, and speciation has occurred here in the $x = 9$ (Sect. *Zarabellia*), $x = 10$ (Sect. *Melampodium*), and $x = 11$ (Sects. *Alcina*, *Glabrata*, *Rhizomaria*) chromosomal lineages. See Chap. 9 for detailed ancestral habitat and area analyses.

Chapter 5
Chromosomal Variation

Among all the fascinating aspects of *Melampodium*, the most impressive is the broad chromosomal variation: $n = 9, 10, 11, 12, 14, 18, 20, 23, 24, 27, 28, 30, 33$. This is not unique among angiosperm genera, e.g., in *Stylidium* (Stylidiaceae) 15 numbers are known from $n = 5$ to 30 (James, 1979), or in *Crepis* (Asteraceae) 24 numbers have been counted from $n = 3$ to 33 (Watanabe, 2008; Semple & Watanabe, 2009; Senderowicz et al., 2021), but such wide variation is not very common. This means, at first instance, that these different numbers in *Melampodium* will likely be significant in the classification and evolution of the group. There are species of some genera that have bizarre infraspecific chromosomal diversity (e.g., 19 different numbers from $2n = 12$ to 72 in *Claytonia virginica*, Portulacaceae, Lewis, 1962, 1970; $2n = 10, 12, 14, 19, 25, 26, 27, 28, 35, 42$ up to 150 in *Prospero autumnale* s.l., Hyacinthaceae, Jang et al., 2018), but these cases are rare. Infrageneric chromosomal change will likely be involved with other factors, such as geography, ecology, or phenology, and with modes of speciation in the genus because differences in chromosome number may cause strong genetic isolation between taxa.

Chromosome Number Reports

An impressive quantity of chromosome numbers in *Melampodium* has been reported in the past 80 years. Most of these are from meiotic squashes, simply because this is the easiest way to get material for cytological data. In members of Asteraceae, the flowering heads develop centripetally, with the outer disc florets developing first. This means that in the production of pollen grains, there will be a sequence of anther stages, some already with mature pollen and others still in the pre-meiotic or meiotic stage. With some experience, one can select small florets with appropriate meiotic stages, squash them in acetocarmine stain, and observe suitable cells. The first report on *Melampodium* was by Negodi (1938), who observed $n = 10$ for *M.*

T. F. Stuessy et al., *Systematics, Ecology, and Chromosomal Evolution of* Melampodium *(Asteraceae)*, https://doi.org/10.1007/978-3-031-81995-7_5

divaricatum and *M. perfoliatum*. These two numbers are not typical for these two species (normally $n = 12$ and 11, respectively), suggesting that the techniques were perhaps imprecise. He also reported $n = 9$ for *M. longifolium*, which is the stable number for this species (Weiss-Schneeweiss et al., 2009). The next reports in the genus came amid a series of studies to document chromosome numbers of various species of plants, especially in Asteraceae (Jackson, 1957; Turner et al., 1961; Turner & Johnston, 1961; Turner & King, 1962; Turner & Flyr, 1966). These provided at least one count each for 21 species of *Melampodium*, which began to reveal broader chromosomal diversity, including the incidence of infraspecific polyploidy in *M. cinereum* and *M. leucanthum* (Turner & King, 1962).

In conjunction with a taxonomic revision of the genus, Stuessy (1970a) summarized previous counts and contributed new ones in the context of a new classification, in the process clarifying which numbers were found in what species, plus adding new counts for *M. glabrum* ($n = 11$). A new count of $n = 12$ for *M. dicoelocarpum* showed the species to contain two cytotypes, the previous count being $n = 23$ (Turner & King, 1962; identified incorrectly as *M. microcephalum*). Of particular interest was the karyotype of *M. sericeum*, $n = 30$, comprising 20 large and 10 small bivalents, suggesting a possible allopolyploid origin for this species. The challenge was to identify the parents involved with such a putative allopolyploid origin, one potentially contributing $n = 20$ and the other $n = 10$. The first chromosomal report of $n = 20$ for the morphologically similar *M. strigosum* (Stuessy, 1970a; confirmed independently by Powell & Sikes, 1970, as *M. hispidum*) suggested that this might be one of the progenitors. What the other diploid parent might have been was uncertain.

The next major synthesis of cytological information in *Melampodium* was that of Stuessy (1971a). He compiled all known chromosome numbers of 26 species totaling 375 plants from 275 populations. One objective was to examine the relationships of chromosomal variation within the genus in the context of the new classification that was based on his thesis work (Stuessy, 1968) and was at that time being prepared for publication (Stuessy, 1972). An important consideration was the stability of the chromosomal level within each taxon. If variation transgresses species and varietal limits, little can be said about relationships based on cytological information. Some species were surveyed intensively, such as *M. leucanthum* (136 plants in 73 populations), *M. divaricatum* (53 plants in 50 populations), *M. americanum* (24 plants in 19 populations), and *M. gracile* (22 plants in 17 populations). Only five species were known from a single report. The results provided confidence in the stability of chromosomal levels within species, with the exceptions of multiple ploidy levels in *M. cinereum* ($n = 10, 20$), *M. dicoelocarpum* ($n = 12, 23$), and *M. leucanthum* ($n = 10, 20$). In this paper, inferences were made on chromosomal evolution within the genus, suggesting that the base number for the genus would most probably be $x = 10$ (see Chap. 7 for more details and discussion).

The most recent chromosomal survey of *Melampodium* was that of Weiss-Schneeweiss et al. (2009). This study presented new mitotic counts for 394 individuals in 111 populations (Table 5.1). Of the original 36 species treated by Stuessy (1972), three new species were added more recently to the genus (*M. mayfieldii*,

Table 5.1 Most recent chromosome count survey among species of *Melampodium* including 394 new (diploid) and all previous (haploid) reports. Figures cited refer to photographs of mitotic counts and idiograms in this book (pp. 64–66). The species are ordered according to the classification of Stuessy (1972). (From Weiss-Schneeweiss et al., 2009)

Taxon	2*n* (number of individuals)	Collection data	Figures	Previous counts for taxon (*n*)
Section *Melampodium*				
Series *Melampodium*				
M. americanum	20 (3)	M, Michoacán; TS, JV, CR, & IC, 18583.	5.7A and 5.9A	10[e,j,k]
	20 (2)	M, Michoacán; TS, JV, CR, & IC, 18592.		
	20 (4)	M, Colima; TS, JV, CR, & IC, 18609.		
	20 (1)	M, Michoacán; TS, JV, CB, & EO, 19009.		
M. diffusum	20 (1)	M, Oaxaca; TS, JV, CR, & IC, 18666.		10[e,j,k]
	20 (4)	M, Guerrero; TS, JV, CR, & IC, 18669.		
	20 (3)	M, Guerrero; TS, JV, CR, & IC, 18671.	5.7B and 5.9B	
	20 (2)	M, Guerrero; TS, JV, CR, & IC, 18672.		
M. linearilobum	20 (5)	M, Oaxaca; TS, JV, CR, & IC, 18661.	5.7C and 5.9C	10[e,j,k]
	20 (2)	M, Oaxaca; TS, JV, CR, & IC, 18662.		
	20 (4)	M, Oaxaca; TS, JV, CR, & IC, 18665.		
	20 (6)	M, Oaxaca; EO, 333, MEXU.		
M. longipes	20 (6)	M, Nayarit; TS, JV, CR, & IC, 18621.	5.7D and 5.9D	10[e,j,k]
M. mayfieldii[o]	40 (1)	M, Colima; TS, JV, CR, & IC, 18613.		–
	39 (1)/40 (2)	M, Jalisco; TS, JV, CB, & EO, 19019.	5.8A	
M. pilosum	20 (4)	M, Michoacán; TS, JV, CR, & IC, 18587.		10[i,k]
	20 (2)	M, Michoacán; TS, JV, CR, & IC, 18589.		
	20 (4)	M, Michoacán; TS, JV, CR, & IC, 18590.	5.7E and 5.9E	
	20 (1)	M, Michoacán; TS, JV, CB, & EO, 19010.		

(continued)

Table 5.1 (continued)

Taxon	2*n* (number of individuals)	Collection data	Figures	Previous counts for taxon (*n*)
Series *Sericea*				
M. longicorne	60 (2)	USA, Arizona, Pima Co.; CR & MB, 18823.	5.8B	30[k]
	60 (4)	USA, Arizona, Pima Co.; CR & MB, 18824.		
M. nayaritense	40 (2)	M, Nayarit; JV, GF & EO, 1575.	5.8C	–
	40 (2)	M, Nayarit; JV, GF & EO, 1577.		
	40 (3)	M, Nayarit; JV, GF & EO, 1579.		
M. pringlei	60 (4)	M, Oaxaca; TS, JV, CR, & IC, 18637.		–
	60 (4)	M, Oaxaca; TS, JV, CR, & IC, 18650.	5.8D	
M. sericeum	60 (4)	M, Michoacán; TS, JV, CR, & IC, 18572.		30[e,j,k]
	60 (7)	M, Oaxaca; TS, JV, CR, & IC, 18620.	5.8E	
M. strigosum[r]	40 (8)	USA, Texas, Jeff Davis Co.; CR & ML, 18728.	5.8F	20[j,k]
	40 (2)	M, Queretaro; TS, JV, & CB, 19073.		
Series *Leucantha*				
M. argophyllum	60 (2)	M, Nuevo León; TS, JV, CR, & CB, 19059.	5.8G	30[q]
	60 (1)	M, Nuevo León; TS, JV, CR, & CB, 19060.		
M. cinereum var. *cinereum*	20 (2)/40 (1)	USA, Texas, Frio Co.; TS & CR, 18688.	5.7F and 5.9F	10, 20[e,g,j,k,q]
	20 (3)	USA, Texas, Dimmit Co.; TS & CR, 18689.		
	40 (3)	USA, Texas, Jim Hogg Co.; TS & CR, 18700.		
	40 (3)/41 (3)	USA, Texas, Jim Hogg Co.; TS & CR, 18703.	5.8H	
M. cinereum var. *hirtellum*	20 (3)	M, Nuevo León; TS, JV, CR, & CB, 19062.		10, 20[j,k,q]
M. cinereum var. *ramosissimum*	20 (1)	M, Tamaulipas; TS, JV, & CB, 19063.		10[j,k,q]
M. leucanthum	20 (10)	USA, Colorado, Freemont co.; SK, WG583 MLK.		10, 20[a,b,e,j,k,q]

(continued)

Table 5.1 (continued)

Taxon	2*n* (number of individuals)	Collection data	Figures	Previous counts for taxon (*n*)
	40 (4)/41 (1)/42 (1) 42 (1)	USA, Texas, Blanco Co.; TS & CR, 18683.	5.8I	
	40 (2)	USA, Texas, Val Verde Co.; CR & ML, 18720.		
	20 (5)	USA, Texas, Reeves Co.; CR & ML, 18727.		
	20 (2)	M, Coahuila; TS, JV, CR, & CB, 19056.	5.7G and 5.9G	
Series *Cupulata*				
M. appendiculatum	20 (4)	M, Sonora; TS, JV, & CB, 19046.	5.7H and 5.9H	10[l]
M. cupulatum	20 (5)	M, Sonora; TS, JV, & CB, 19044.	5.7I and 5.9I	10[g,k] 12II + 1[p]
	20 (3)	M, Sonora; TS, JV, & CB, 19045.		
	20 (2)	M, Sonora; TS, JV, & CB, 19047.		
M. glabribracteatum	20 (8)	M, Oaxaca; TS, JV, CR, & IC, 18654.	5.7J and 5.9J	–
M. moctezumum	2*x***	M, Sonora; TD & AR, 2003–1228, TEX.		–
M. rosei	20 (3)	M, Sinaloa; TS, JV, CB, & EO, 19029.		10[e,g,j,k]
	20 (2)	M, Sinaloa; TS, JV, CB, & EO, 19032.		
	20 (2)	M, Sinaloa; TS, JV, CB, & EO, 19040.		
	20 (3)	M, Sinaloa; TS, JV, & CB, 19050.	5.7K and 5.9K	
M. sinuatum	**n* = 10 (1)	M, Baja California; TS & JV, 19037.	5.7K	–
M. tenellum	20 (4)	M, Nayarit; TS, JV, CB, & EO, 19020.	5.7M and 5.9L	10[e,j,k]
	20 (2)	M, Nayarit; TS, JV, CB, & EO, 19022.		
Series *Longipila*				
M. longipilum	20 (10)	M, Oaxaca; TS, JV, CR, & IC, 18627.	5.7N and 5.9M	10[j,k], 11[e]
	20 (2)	M, Oaxaca; TS, JV, CR, & IC, 18630.		
	20 (5)	M, Oaxaca; TS, JV, CR, & IC, 18648.		

(continued)

Table 5.1 (continued)

Taxon	2*n* (number of individuals)	Collection data	Figures	Previous counts for taxon (*n*)
	20 (3)	M, Puebla; EO, 327, MEXU.		
	20 (4)	M, Oaxaca; EO, 332, MEXU.		
	20 (1)	M, Oaxaca; EO, 342, MEXU.		
Section *Zarabellia*				
M. gracile	18 (1)	M, Michoacán; TS, JV, CR, & IC, 18586.		9[d,e,g,j,k]
	18 (1)/19 (1)	M, Colima; TS, JV, CR, & IC, 18606.		
	18 (1)	M, Tamaulipas; TS, JV, & CB, 19066.		
	18 (3)	M, Tamaulipas; TS, JV, & CB, 19067.	5.7O and 5.9N	
	18 (2)	M, Tamaulipas; TS, JV, & CB, 19068.		
M. longifolium	18 (2)	M, Oaxaca; TS, JV, CR, & IC, 18629.	5.7Q1 and 5.9P	9[a,g,j,k]
	18 (3)	M, Oaxaca; TS, JV, CR, & IC, 18633.		
	18 (2)	M, Queretaro; TS, JV, & CB, 19071.		
	18 (3)	M, Mexico D.F.; TS, JV, & CB, 19074.	5.7Q2 and 5.9Q	
M. microcephalum	18 (3)	M, Michoacán; TS, JV, CR, & IC, 18574.		9[e,j,k]
	18 (7)	M, Oaxaca; TS, JV, CR, & IC, 18651.	5.7P and 5.9O	
	18 (4)	M, Oaxaca; EO, 334, MEXU.		
M. mimulifolium	18 (5)	M, Oaxaca; TS, JV, CR, & IC, 18656.	5.7R and 5.9R	–
M. paniculatum	36 (3)	M, Chiapas; JV, EO, & JC, 1589.	5.8J	18[e,j,k], 27[h,i,k]
	36 (6)	M, Chiapas; JV, EO, & JC, 1591.		
	36 (4)	M, Chiapas; JV, EO, & JC, 1593.		
Section *Serratura*				
M. costaricense	48 (4)	CR, Prov. San José; TS, JV, & IS, 19079.		25 ± 1[f,k]

(continued)

Table 5.1 (continued)

Taxon	2*n* (number of individuals)	Collection data	Figures	Previous counts for taxon (*n*)
	48 (5)	CR, Prov. San José; TS, JV, & IS, 19080.	5.8K	
M. dicoelocarpum	24 (2)/46 (2)	M, Michoacán; TS, JV, CR, & IC, 18588.	5.7S and 5.9S	12[j,k], 23[e,j,k]
	46 (2)	M, Michoacán; TS, JV, CR, & IC, 18595.	5.8L	
	46 (1)	M, Michoacán; TS, JV, CR, & IC, 18599.		
	46 (1)	M, Jalisco; TS, JV, CR, & IC, 18603.		
M. divaricatum	24 (4)	M, Michoacán; TS, JV, CR, & IC, 18573.	5.7T and 5.9T	10[a], 12[e,j,k], c. 30II[p]
	24 (3)	CR, Prov. San José; TS, JV, & IS, 19086.		
	24 (8)/ 24 + 1B (1)	M, Oaxaca; TS, JV, CR, & IC, 18668.		
	24 (2)	M, Guerrero; TS, JV, CR, & IC, 18670.		
	24 (2)	M, San Luis Potosí; TS, JV, & CB, 19069.		
M. northingtonii[n]	48 (7)	M, Oaxaca; TS, JV, CR, & IC, 18659.	5.8M	–
M. sinaloense	48 (2)	M, Sinaloa; TS, JV, CB, & EO, 19026.	5.9N	–
	48 (1)	M, Sinaloa; TS, JV, CB, & EO, 19027.		
M. tepicense	24 (4)	M, Nayarit; TS, JV, CR, & IC, 18614.	5.7U and 5.9U	12[m]
	24 (6)	M, Nayarit; TS, JV, CR, & IC, 18615.		
Section *Bibractiaria*				
M. bibracteatum	56 (2)	M, México; TS, JV, CB, & EO, 19002.	5.8O	–
	56 (2)	M, Durango; TS, JV, CR, & IC, 19052.		
M. repens	28 (1)	M, México; TS, JV, CB, & EO, 19001.	5.7V and 5.9V	27[m]
	28 (1)	M, Durango; TS, JV, CR, & CB, 19053.		
Section *Rhizomaria*				
M. aureum	66 (6)	M, Michoacán; TS, JV, CR, & IC, 18576.	5.8P	33[e,j,k]

(continued)

Table 5.1 (continued)

Taxon	2*n* (number of individuals)	Collection data	Figures	Previous counts for taxon (*n*)
	66 (3)	M, Michoacán; TS, JV, CR, & IC, 18580.		
	66 (2)	M, Michoacán; TS, JV, CR, & IC, 18582.		
	66 (2)	M, Oaxaca; TS, JV, CR, & IC, 18635.		
M. montanum var. *montanum*	**n* = 11 (1)	M, Oaxaca; TS, JV, CR, & IC, 18640.		ca. 11[e,k]
M. montanum var. *viridulum*	22 (10)/ 22 + 5Bs (1)	M, Oaxaca; TS, JV, CR, & IC, 18646.	5.7W and 5.9W	11[c,e,j,k]
	22 (6)	M, Oaxaca; TS, JV, CR, & IC, 18655.		
Section *Alcina*				
M. glabrum	22 (5)	M, Michoacán; TS, JV, CR, & IC, 18624.	5.7X and 5.9X	11[j,k]
	22 (4)	M, Michoacán; TS, JV, CR, & IC, 19012.		
M. nutans	22 (6)	M, Michoacán; TS, JV, CR, & IC, 18591.		11[i,k]
	22 (5)	M, Colima; TS, JV, CR, & IC, 18611.		
	22 (7)	M, Oaxaca; TS, JV, CR, & IC, 18664.	5.7Y and 5.9Y	
M. perfoliatum	22 (10)	M, Michoacán; TS, JV, CR, & IC, 18568.		10[a], 11[e,f,j,k], 12[e,f,k]
	22 (2)	M, Oaxaca; TS, JV, CR, & IC, 18652.	5.7Z and 5.9Z	

All vouchers are deposited in WU unless otherwise indicated. Countries: *CR* Costa Rica, *E* Ecuador, *M* México, *USA* United States of America. Collectors: *AR* A.L. Reina, *CB* C. Blöch, *CR* C.A. Rebernig, *EO* E. Ortiz B., *IC* I. Calzada, *IS* I. Sánchez, *JV* J.L. Villaseñor, *MB* M.H.J. Barfuss, *ML* M. Lenko, *JC* J. Calónico, *SK* S. Kelso, *TD* T.R. Van Devender, *TS*, T.F. Stuessy. All chromosome numbers obtained from root tip meristems, unless otherwise indicated (*meiosis in pollen mother cells; **analyzed using flow cytometry of mature seeds, J. Rauchova and H. Weiss-Schneeweiss, unpubl.)

[a]Negodi (1938)
[b]Jackson (1957)
[c]Turner et al. (1961)
[d]Turner and Johnston (1961)
[e]Turner and King (1962)
[f]Turner and King (1964)
[g]Turner and Flyr (1966)
[h]Coleman (1970)

(continued)

Table 5.1 (continued)

[i]Stuessy (1970a)
[j]Stuessy (1971a)
[k]Stuessy (1971b)
[l]Keil and Stuessy (1975)
[m]Keil and Stuessy (1977)
[n]Turner (1988)
[o]Turner (1993)
[p]Carr et al. (1999)
[q]Stuessy et al. (2004)
[r]Published as *M. hispidum*

M. moctezumum, *M. northingtonii*; Turner, 1988, 1993, 2007, respectively). Table 5.1 lists the species organized by the classification of Stuessy (1972) with known chromosome numbers from previous studies and the addition of many new counts. Emphasis was also on analyzing as many individuals as practicable within populations to determine uniformity within a species or to detect infraspecific variation.

Weiss-Schneeweiss et al. (2009) contributed first reports for 11 species of *Melampodium*. These were from *M. bibracteatum* ($2n = 56$), *M. glabribracteatum* ($2n = 20$), *M. mayfieldii* ($2n = 40$, with one individual $2n = 39$), *M. mimulifolium* ($2n = 18$), *M. moctezumum* ($2x$, interpreted in phylogenetic context as $2n = 20$, as determined by flow cytometry of mature seeds), *M. nayaritense* ($2n = 40$), *M. northingtonii* ($2n = 48$), *M. pringlei* ($2n = 60$), *M. repens* ($2n = 28$), *M. sinaloense* ($2n = 48$), and *M. sinuatum* ($2n = 20$). Particularly significant were the reports from Sect. *Bibractiaria* (*M. repens* and *M. bibracteatum*) with $2n = 28$ and 56, respectively, indicating a diploid-tetraploid relationship on a new base of $x = 14$, further stressing the extensive cytological variation within the genus. One previous count for *M. repens* was reported as $n = 27$ by Keil and Stuessy (1977), presumably a miscount. The count for *M. costaricense* was also clarified to be $2n = 48$ (previously known as $n = 25 \pm 1$; Turner & King, 1964 [reported as *M.* cf. *flaccidum*]; Stuessy, 1971a), which was interpreted as a tetraploid in comparison with the very close relative, *M. divaricatum*, a diploid at $2n = 24$.

A positive contribution from this enhanced chromosomal survey was revealing the extensive occurrence of polyploidy among species of the genus. Thirteen species were seen to be exclusively polyploid (seven tetraploid, five hexaploid, and one with both these ploidy levels), and four with diploid and tetraploid cytological races (*M. dicoelocarpum* containing diploids and fixed dysploidy at the polyploid level, $n = 12$ and 23). Autopolyploid origins were suggested for the higher ploidy levels in *M. cinereum* and *M. leucanthum* ($n = 20$), and possibly also *M. aureum* ($n = 33$). Allopolyploidy was also proposed for the origin of *M. paniculatum* ($n = 18$), *M. argophyllum* ($n = 30$), *M. nayaritense* ($n = 20$; 10 large and 10 smaller bivalents), and *M. pringlei* and *M. sericeum* (both with 20 large and 10 small bivalents).

White-Rayed Complex

Due to the existence of infraspecific polyploidy in the white-rayed *Melampodium cinereum* and *M. leucanthum*, more populational surveying was carried out to determine the geographic distributions of the cytotypes. Mixed-ploidy plant species are not uncommon in Asteraceae (Kolář et al., 2017), and their existence in the white-rayed complex of *Melampodium* was detected early by Turner and King (1962). It is important to stress that these cytological variants were never considered to be anything other than autopolyploid cytotypes, in contrast to being accorded formal status as varieties, subspecies, or even forms. Since reproductive crossing between different cytological levels is highly restricted, some previous workers (e.g., Löve et al., 1957; Löve & Löve, 1964) recommended treating different cytotypes as good biological species. In these two white-rayed species of *Melampodium*, however, there is no qualitative morphological difference between the cytotypes, which means limited genetic difference between them, hence vitiating specific recognition.

Stuessy (1971b) surveyed the geographic distributions of diploid and tetraploid cytotypes in both *M. cinereum* and *M. leucanthum*, with a much larger sample in the latter because of its broad distribution from Arizona to central Texas and northward through New Mexico to Colorado. The results (Fig. 5.1) showed that patterns existed within both species with tetraploids being largely confined to the easternmost parts of the ranges. In *M. cinereum*, which occurs in the sandy plains south of the Edwards Plateau in central Texas, south to the Rio Grande River, and into northern Mexico, the tetraploids formed a contiguous distribution.

Because the tetraploid individuals in *M. cinereum* var. *cinereum* were somewhat larger than the diploids, it was possible to map the distribution of these morphological forms to give a more detailed picture of the pattern of distribution of the cytological races (Fig. 5.2). The same basic pattern holds, except that more tetraploids are seen along the Rio Grande River, and some diploids are intermixed among the tetraploids. In *M. leucanthum* diploids prevail throughout the range except on the calcareous Edwards Plateau, where tetraploids predominate. The region around Austin, Texas, is especially interesting because here diploids and tetraploids co-occur in Blanco and Travis Counties.

More intensive cytological surveying on the white-rayed complex was published by Stuessy et al. (2004). The objective was to add many more populational counts to test the geographic patterns documented by Stuessy (1971b). New data were analyzed from 415 plants in 152 populations (Fig. 5.3), which when added to existing information yielded a total of 600 individuals surveyed from 152 populations. Within *M. leucanthum*, combined data revealed a pattern similar to what had been presented earlier, with the tetraploids being largely confined to the eastern part of the range of the species, but now with a few exceptions. Three populations along the western edge of the Texas panhandle were encountered with both diploid and tetraploid individuals. Likewise, in Blanco Co. (Fig. 5.3 inset) two mixed populations occur plus one triploid plant. More intensive intrapopulational surveying was also

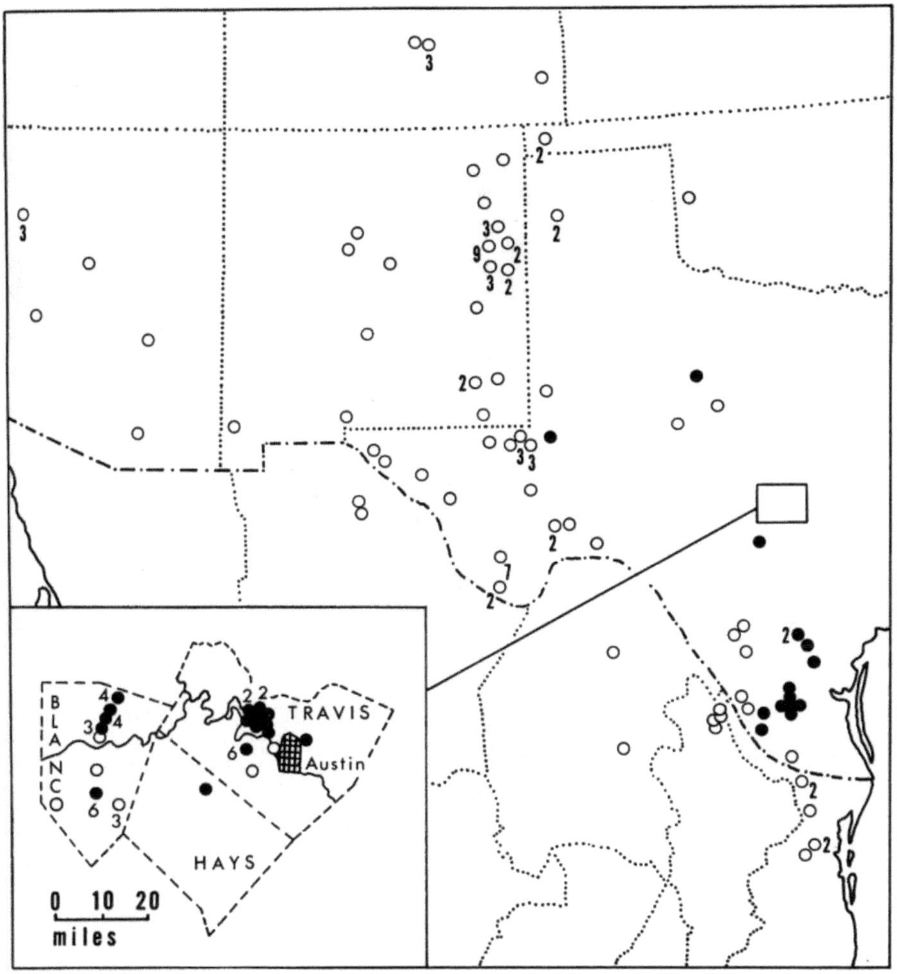

Fig. 5.1 Distribution of chromosome counts in populations of *Melampodium leucanthum* and *M. cinereum* (open circles, $n = 10$; closed circles, $n = 20$). Numbers indicate the number of plants analyzed in each population (if more than one). (From Stuessy, 1971b: 183)

completed, particularly the 27-plant sample from a population in the Big Bend region of Texas (all diploid).

An intensive survey of populations within the three morphological varieties of *M. cinereum* was also conducted (Stuessy et al., 2004), including analysis of more individuals within populations (Fig. 5.4). This species consists of three morphological varieties, *M. cinereum* var. *cinereum*, var. *hirtellum*, and var. *ramossisimum*. The latter variety occurs in northern Mexico (Tamaulipas) and is characterized by very tomentose leaves. Variety *hirtellum* occurs primarily in Nuevo León and Coahuila, Mexico, with occasional individuals further north into Texas, and it is distinguished by hirtellous stems. The southern Texas populations are primarily of var. *cinereum*.

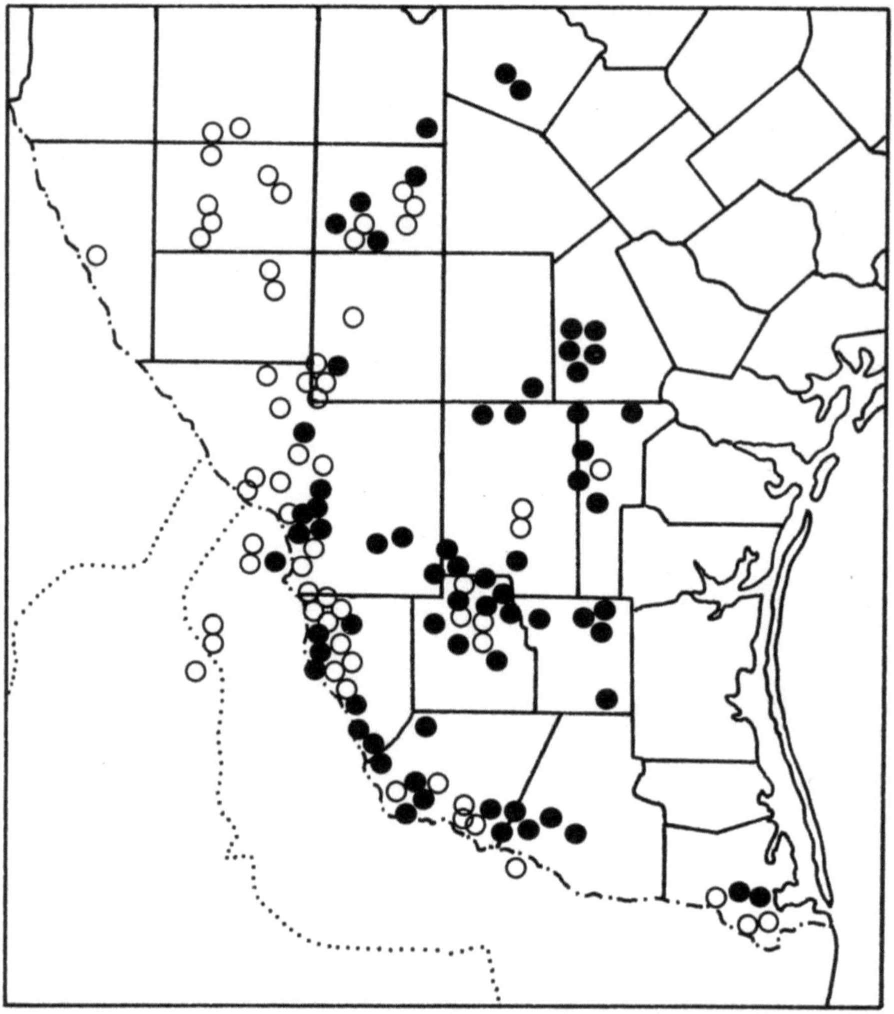

Fig. 5.2 Distribution of the morphological forms of *Melampodium cinereum* var. *cinereum* in southern Texas and adjacent Mexico (closed circles, inferred $n = 20$; open circles, inferred $n = 10$). (From Stuessy, 1971b: 187)

The geographic pattern in var. *cinereum* stayed approximately the same as reported previously, and in fact, the actual cytological data were not too removed from those from inferred ploidy levels based on quantitative aspects of morphology (cf. Fig. 5.2). These more intensive surveys showed that mixed populations of diploids and tetraploids also occur, albeit rarely, in var. *ramosissimum* and var. *hirtellum*.

Another interesting result of these studies on the white-rayed complex was new cytological data for *M. argophyllum*. This species was originally proposed by Asa Gray but formally described by Robinson (1901) as a variety of *M. cinereum*, but later it was treated as a good species by S. F. Blake (1924) and for a brief period as

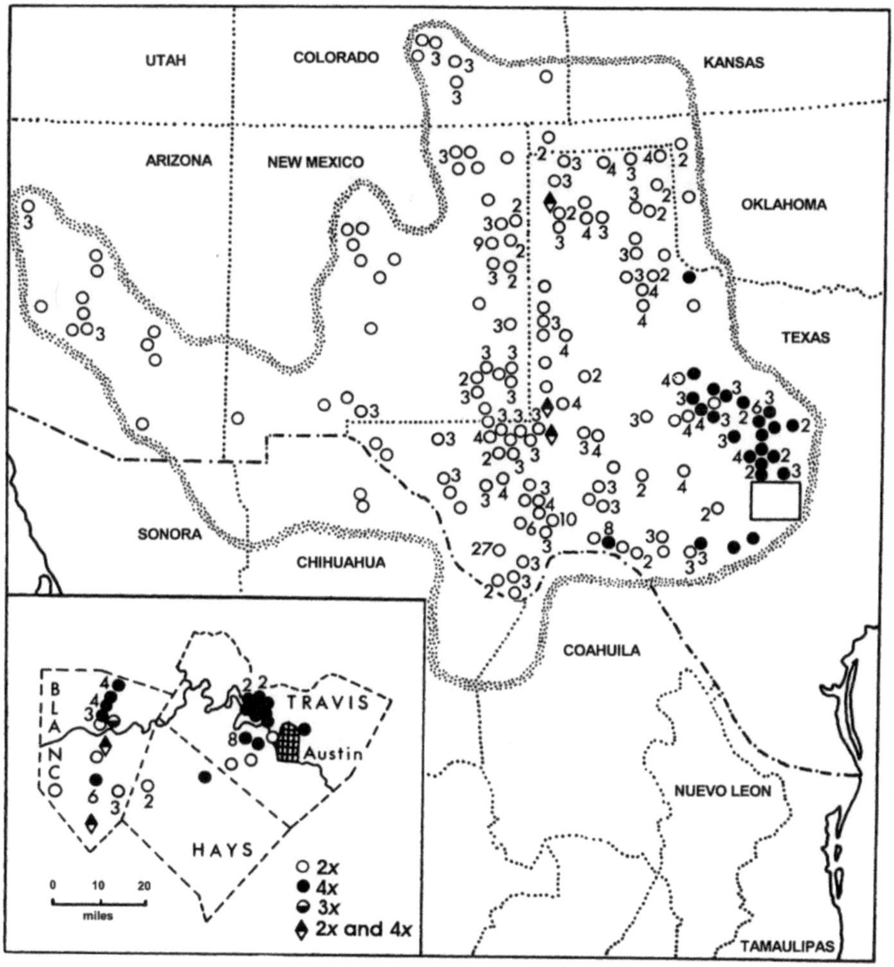

Fig. 5.3 Distribution of chromosome counts in populations of *Melampodium leucanthum*. Numbers indicate individuals counted in each population (if more than one). (From Stuessy et al., 2004: 890)

a variety of *M. leucanthum* by Stuessy (1969b). First reports for this species by Stuessy et al. (2004) revealed it to be hexaploid ($n = 30$). This surprising result gave support to its recognition as a good species, but it also raised questions about its evolutionary origin. So far, no diploid populations of this taxon have been found, but the species is not common. Based on what was known at the time, one would expect an allopolyploid origin of *M. argophyllum* from diploid and tetraploid ancestors from within the rest of the white-rayed complex.

A more recent analysis of cytological races in the white-rayed complex, presented for the first time here, has been an offshoot of AFLP studies on the group (Rebernig et al., 2010a, b, 2012). The approach used flow cytometry to determine

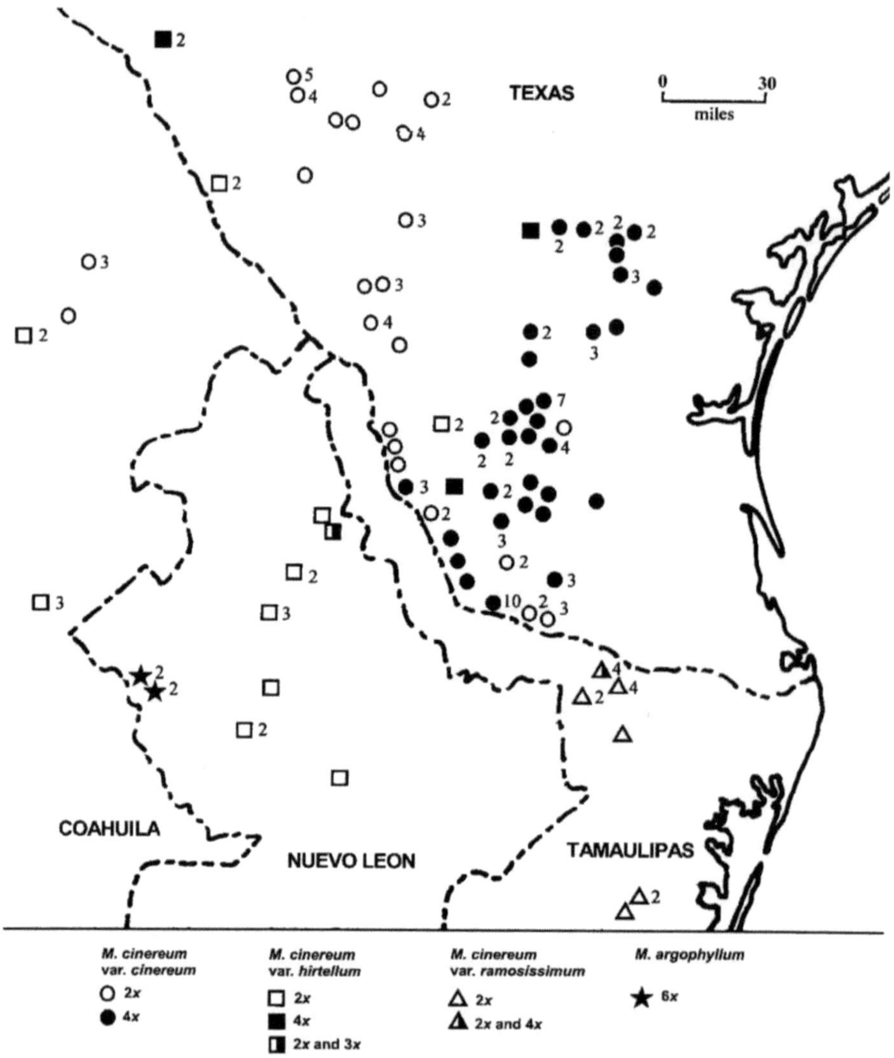

Fig. 5.4 Distribution in southern Texas and northeastern Mexico of chromosome counts in populations of *Melampodium cinereum* (2*x*, 3*x*, 4*x*) and *M. argophyllum* (6*x*, *n* = 30). (From Stuessy et al., 2004: 891)

ploidy levels (relative amounts of DNA) from silica-dried leaves (e.g., Doležel et al., 2007; Kron et al., 2007). This method is sufficiently sensitive to distinguish among 2*x*, 3*x*, 4*x*, 5*x*, and 6*x* individuals, with results being calibrated by comparisons with genome sizes of plants of the same species counted chromosomally. The investigation provides data on ploidy levels for a total of 2722 plants in 191 populations of the white-rayed complex: 2239 plants in 159 populations of *M. leucanthum*, 448 plants in 30 populations of *M. cinereum*, and 35 plants in 2 populations of *M. argophyllum* (Table 5.2; Appendix 1).

Table 5.2 Ploidy levels of plants in the white-rayed complex of *Melampodium* as determined by flow cytometry

Taxon	Number of populations	Number of individuals	Ploidy level				
			2*x*	3*x*	4*x*	5*x*	6*x*
M. argophyllum	2	35					35
M. cinereum							
var. *cinereum*	20	270	36		229	4	1
var. *hirtellum*	6	138	137	1			
var. *ramosissimum*	4	40	40				
Subtotals	30	448	213	1	229	4	1
M. leucanthum							
Arizona	58	993	992	1			
New Mexico	1	29	29				
Texas	100	1217	800	10	406		1
Subtotals	159	2239	1821	11	406		1
Grand totals	191	2722	2034	12	635	4	37

See populational data in Appendix 1. Leaves were collected in silica gel and processed in the laboratory by Renate Obermayer following protocols in Doležel et al. (2007)

This new survey reveals more cytological diversity than known previously. We now see more triploids, mixed diploid and tetraploid populations, and occasional pentaploid and hexaploid individuals. Most of the populations are pervasively diploid or tetraploid, only rarely containing triploids, pentaploids, or hexaploids. The data for 35 plants in two populations of *M. argophyllum* continue to reveal only 6*x* plants. This is the only species in the white-rayed complex to be cytologically uniform, at least as far as presently known. The taxon grows in isolated and difficult to access regions in the mountains of western Nuevo León and southeastern Coahuila (Fig. 5.4), which makes intensive sampling challenging.

The results of investigations on populations of *M. cinereum* are best discussed for each of the three morphological varieties, var. *cinereum*, var. *hirtellum*, and var. *ramosissimum* (Table 5.2; see also Appendix 1). Previous counts by Rebernig et al. (2012) for *M. cinereum* came from 16 plants in 16 populations of var. *cinereum*, 6 plants in 6 populations of var. *hirtellum*, and 2 plants in 2 populations of var. *ramosissimum*. The new data from flow cytometry show that var. *ramosissimum* from a sample of 40 plants in 4 populations is exclusively diploid. Variety *hirtellum* is diploid, with only one triploid individual encountered in a population (*Stuessy et al. 19058*) intermixed with 27 diploid plants. Within var. *cinereum*, 13 populations are 4*x* and seven are 2*x*, and in the latter only diploids occur. Within the 4*x* populations, pentaploid and hexaploid individuals have been sporadically encountered, but only five of a total of 270 plants. In populations *18694*, one 6*x* plant occurred, and in *18698*, *18701*, and *18708*, 1, 1, and 2 pentaploids were found, respectively (see Appendix 1).

Previous ploidy level determinations by flow cytometry of 377 plants in 93 populations of *M. leucanthum* were published by Rebernig et al. (2010b) and of 92 plants in the same number of populations by Rebernig et al. (2012). The new data (Table 5.2; Appendix 1) for *M. leucanthum* presented in this book for 2239 plants in 159 populations greatly extend this survey and reveal that triploids are extremely rare in all populations, only showing up more conspicuously in eastern Texas where the concentration of tetraploids is high. One triploid individual has been documented in population *20018* from Arizona with 21 plants sampled in which the other 20 individuals were diploid. Likewise, one triploid plant was found in population *20039* from western Texas (near Big Bend National Park) among 14 diploid plants. Triploids have also been encountered in populations of eastern Texas that are mostly 4*x*, a total of 9 triploid plants and 406 tetraploid individuals. Only one hexaploid was discovered in population *18780* associated with 14 tetraploids. This new survey has revealed that most of the populations are pervasively diploid or tetraploid, only rarely containing triploid, pentaploid, or hexaploid individuals, but it also emphasizes that these two species tend to polyploidize.

Geographic plotting of the new data of cytotypes (Fig. 5.5) of *M. leucanthum* and *M. cinereum* yields the same pattern as seen in Fig. 5.3, in which the tetraploids in both these species are concentrated in the eastern portions of their ranges, either on the Edward's Plateau or in the southern adjacent sandy plains, respectively. Of interest is that this pattern is very similar to that from cytological surveying from

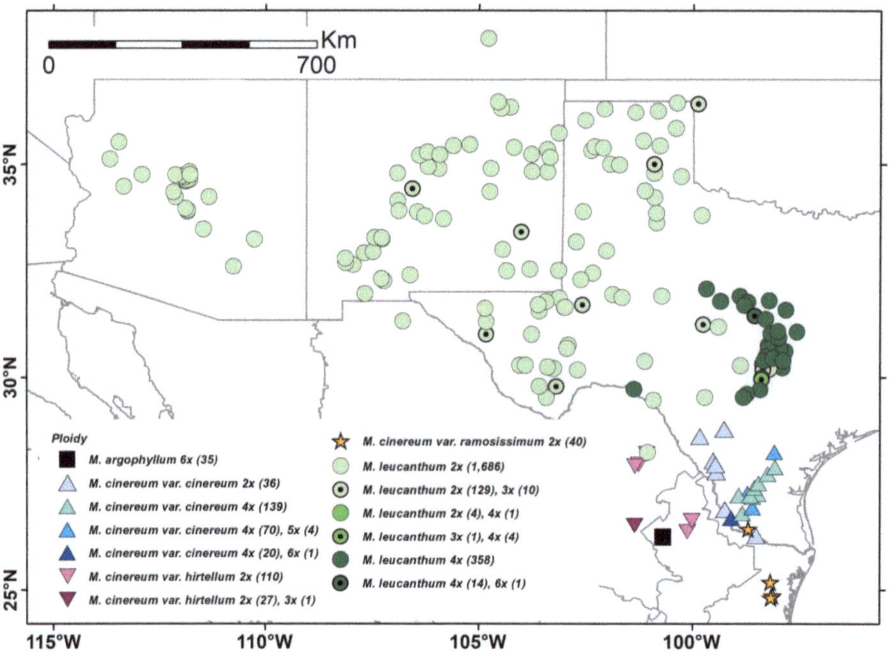

Fig. 5.5 Distribution of ploidy levels in the white-rayed species complex of *Melampodium* as determined by flow cytometry. (Data from Renate Obermayer)

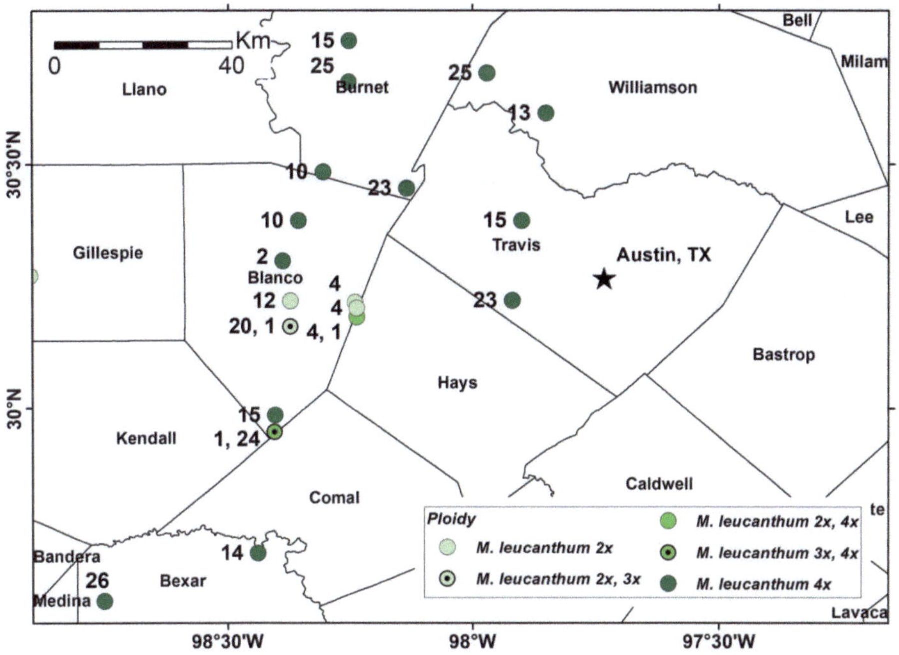

Fig. 5.6 Ploidy level variation within and among populations of *Melampodium leucanthum* in counties near Austin, Texas. Numbers show the total individuals analyzed in each population. (Based on flow cytometry data from Renate Obermayer)

collections made in the late 1960s and early 1970s (Stuessy, 1971b; Stuessy et al., 2004; Figs. 5.1 and 5.3). The geographic pattern of tetraploidy, therefore, appears stable over 50 years. More mixed ploidy populations are now documented, especially in the counties near Austin, Texas (Fig. 5.6). The life span of the species in the white-rayed complex has not been determined with certainly, but they are subshrubs with several annual rings in the stems at the base. This suggests that the pattern of cytological races would be stable over approximately 10 generations. What the environmental conditions might be that selected for tetraploidy in the eastern portions of the ranges are unknown.

Karyotypes

The mitotic chromosomes of *Melampodium* (Figs. 5.7 and 5.8) are medium-sized (the smallest 0.85–1.47 μm in *M. linearilobum*; the largest 2.08–2.92 μm in *M. repens*), and serve, with detailed karyotypic study (Fig. 5.9) as a prelude to understanding mechanisms of chromosomal evolution within the genus. There is no evidence of simple fusions or fissions of chromosomes, although dysploidy has occurred during evolution of the group. There is also ample evidence of numerous

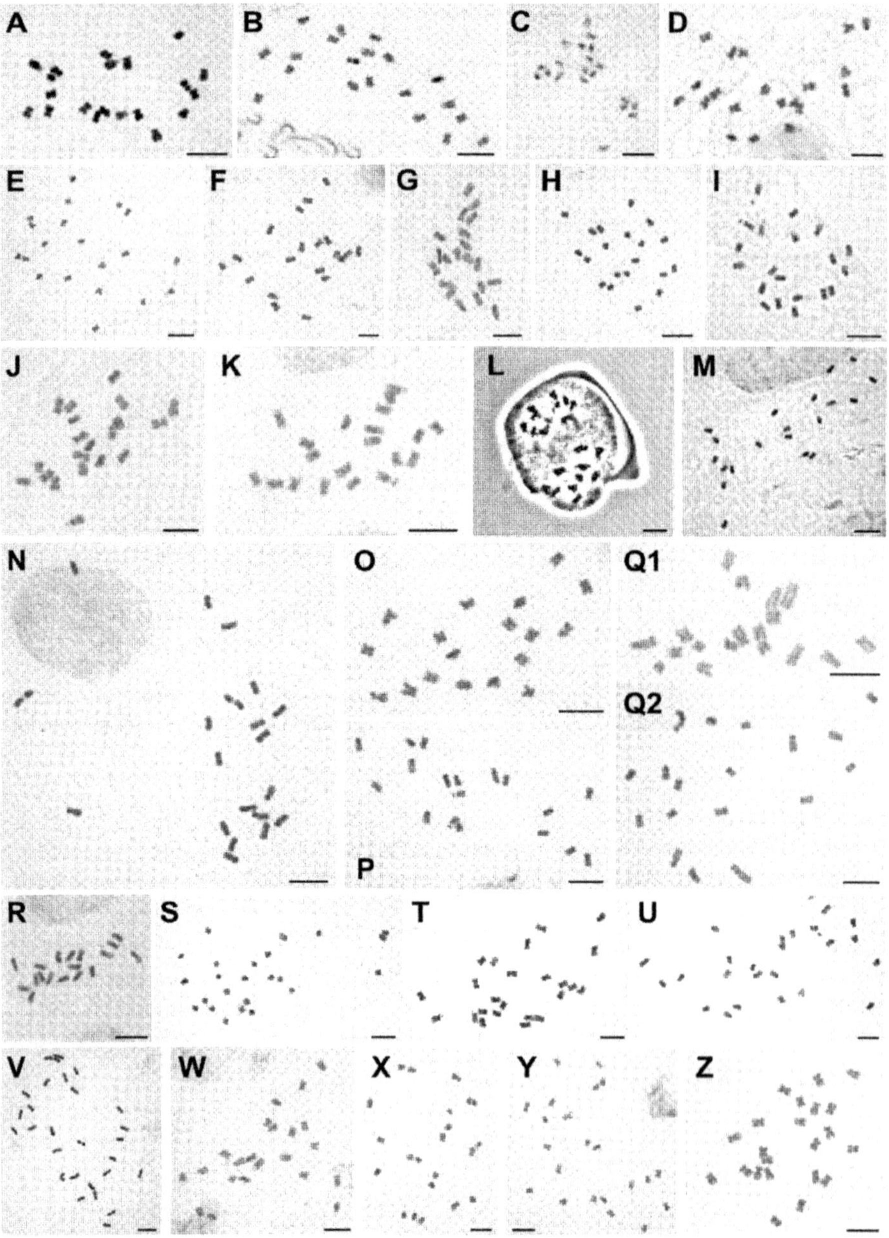

Fig. 5.7 Chromosomes of diploid species of *Melampodium*. (**A–N**) all $2n = 2x = 20$. (**A**) *M. americanum*; (**B**) *M. diffusum*; (**C**) *M. linearilobum*; (**D**) *M. longipes*; (**E**) *M. pilosum*; (**F**) *M. cinereum* var. *cinereum*; (**G**) *M. leucanthum*; (**H**) *M. appendiculatum*; (**I**) *M. cupulatum*; (**J**) *M. glabribracteatum*; (**K**) *M. rosei*; (**L**) *M. sinuatum* (meiotic count); (**M**) *M. tenellum*; (**N**) *M. longipilum*. (**O–R**) all $2n = 2x = 18$. (**O**) *M. gracile*; (**P**) *M. microcephalum*; (**Q1, Q2**) *M. longifolium*; (**R**) *M. mimulifolium*. (**S–U**) all $2n = 2x = 24$. (**S**) *M. dicoelocarpum*; (**T**) *M. divaricatum*; (**U**) *M. tepicense*. (**V**) $2n = 2x = 28$, *M. repens*. (**W–Z**) all $2n = 2x = 22$. (**W**) *M. montanum* var. *viridulum*; (**X**) *M. glabrum*; (**Y**) *M. nutans*; (**Z**) *M. perfoliatum*. Scale bars = 5 μm. (From Weiss-Schneeweiss et al., 2009: 1172)

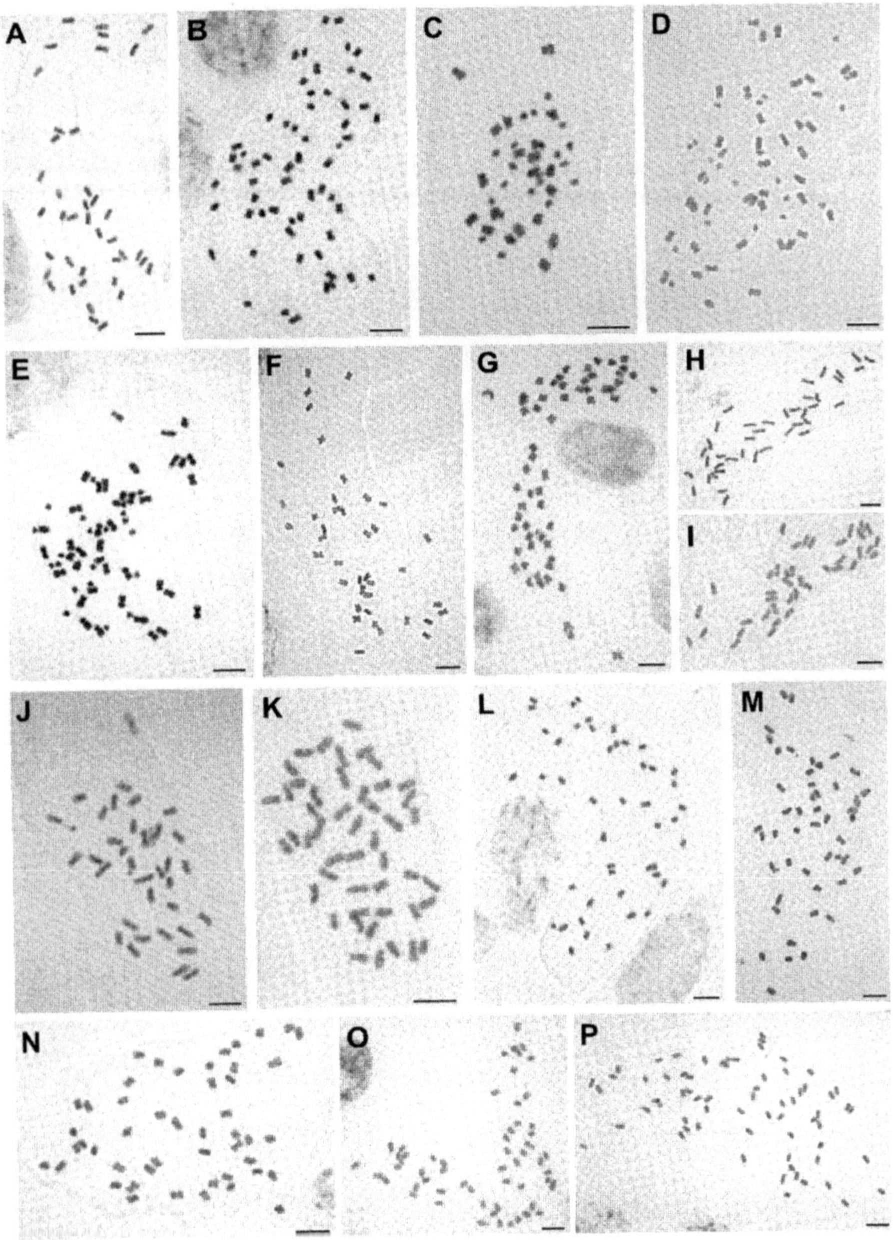

Fig. 5.8 Chromosomes of polyploid species of *Melampodium*. (**A**) *M. mayfieldii* ($2n = 4x = 40$); (**B**) *M. longicorne* ($2n = 6x = 60$); (**C**) *M. nayaritense* ($2n = 4x = 40$); (**D**) *M. pringlei* ($2n = 6x = 60$); (**E**) *M. sericeum* ($2n = 6x = 60$); (**F**) *M. strigosum* ($2n = 4x = 40$); (**G**) *M. argophyllum* ($2n = 6x = 60$); (**H**) *M. cinereum* var. *cinereum* ($2n = 4x = 40$); (**I**) *M. leucanthum* ($2n = 4x = 40$); (**J**) *M. paniculatum* ($2n = 4x = 36$); (**K**) *M. costaricense* ($2n = 4x = 48$); (**L**) *M. dicoelocarpum* ($2n = 4x\text{-}2 = 46$); (**M**) *M. northingtonii* ($2n = 4x = 48$); (**N**) *M. sinaloense* ($2n = 4x = 48$); (**O**) *M. bibracteatum* ($2n = 4x = 56$); (**P**) *M. aureum* ($2n = 6x = 66$). Scale bars = 5 μm. (From Weiss-Schneeweiss et al., 2009: 1174)

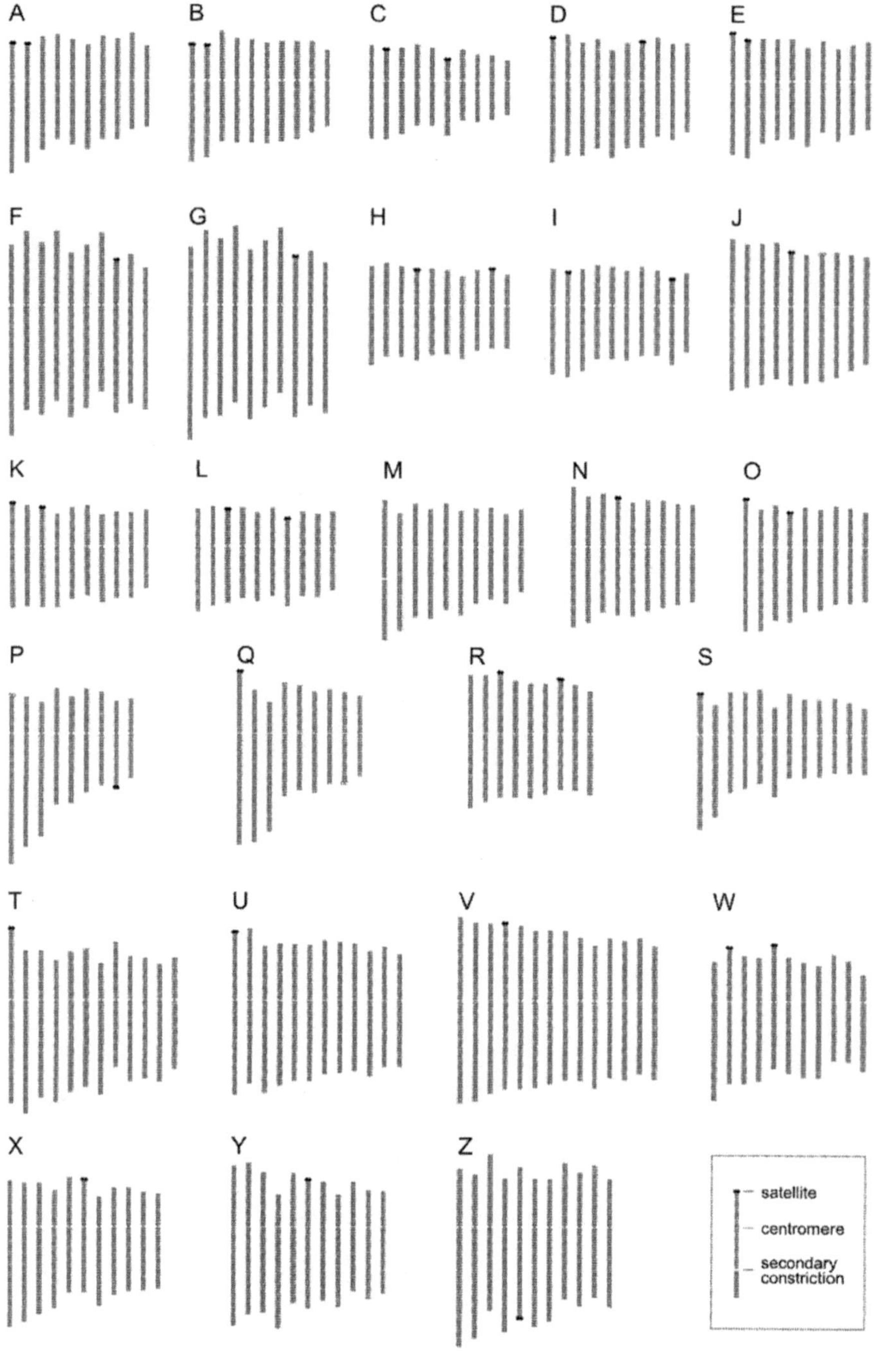

Fig. 5.9 Idiograms of diploid species of *Melampodium*. (**A**) *M. americanum*; (**B**) *M. diffusum*; (**C**) *M. linearilobum*; (**D**) *M. longipes*; (**E**) *M. pilosum*; (**F**) *M. cinereum* var. *cinereum*;

polyploidy events within five different chromosomal lineages: (1) $n = 10, 20, 30$; (2) 9, 18, 27; (3) 11, 33; (4) 12, 24 (plus 23); (5) 14, 28.

In Sect. *Melampodium*, with all species based on $x = 10$, among the diploid species of Ser. *Melampodium* the karyotypes are similar, with mostly meta- and submetacentric chromosomes and two chromosome pairs with subterminal nucleolus organizer regions (NORs). Of interest is that *M. linearilobum* has much smaller chromosomes (Fig. 5.7C and 5.9C), which will become important in discussions of the origin of allopolyploids (see Chap. 7). The species of Ser. *Sericea* are all polyploids with *M. strigosum* and *M. nayaritense* being tetraploids ($2n = 40$) and *M. longicorne*, *M. pringlei*, and *M. sericeum* being hexaploids ($2n = 60$). Karyotypes of *M. nayaritense*, *M. pringlei*, and *M. sericeum* are bimodal with 20 large and 20 small chromosomes in the former and 40 large and 20 small in the latter two. *M. strigosum* is essentially unimodal and *M. longicorne* is slightly bimodal. All species of this series possess two chromosome pairs with subterminal NOR, except for *M. pringlei* with three (possibly four) pairs. In the white-rayed complex, the diploids of *M. cinereum* and *M. leucanthum* have symmetrical and unimodal karyotypes. The chromosomes are relatively large with one pair containing a NOR. In Ser. *Longipila*, the diploid *M. longipilum* contains a secondary constriction, a NOR (Fig. 5.9M) located interstitially in the pericentromeric region of chromosome 1.

Within Sect. *Zarabellia*, with $x = 9$, the very closely related *M. gracile* and *M. microcephalum* have very similar karyotypes. These two species are closely related to the tetraploid *M. paniculatum* ($n = 18$; some hexaploid counts are also known, $n = 27$; Stuessy & Brunken, 1979). The more distantly related *M. longifolium* of the same section is also diploid with $n = 9$, has a more differentiated and asymmetrical karyotype (Fig. 5.9P, Q), and shows some infraspecific variation.

In Sect. *Serratura* the chromosome numbers are based on $x = 12$. Both *M. divaricatum* and *M. tepicense* are both exclusively diploid with $2n = 24$. The other species of the section, *M. costaricense*, *M. northingtonii*, and *M. sinaloense*, are uniformly tetraploid with $2n = 48$. *Melampodium dicoelocarpum* is interesting in that two cytotypes exist within the species, a diploid at $2n = 24$ (Fig. 5.7S) and a dysploid at the tetraploid level, $2n = 46$. These forms have been found within the same population (*Stuessy et al. 18588*), and they cannot be differentiated morphologically. As in other sections of the genus, some related species have similar karyotypes (e.g., *M. divaricatum* and *M. tepicense*, both symmetrical), and others are different, such as *M. dicoelocarpum* with an asymmetric karyotype and smaller chromosomes.

The recent studies by Weiss-Schneeweiss et al. (2009) have revealed a new base number for *Melampodium*, $x = 14$, in Sect. *Bibractiaria*. With all the previous existing chromosomal diversity within the genus, it was surprising that another

Fig. 5.9 (continued) (**G**) *M. leucanthum*; (**H**) *M. appendiculatum*; (**I**) *M. cupulatum*; (**J**) *M. glabribracteatum*; (**K**) *M. rosei*; (**L**) *M. tenellum*; (**M**) *M. longipilum*; (**N**) *M. gracile*; (**O**) *M. microcephalum*; (**P, Q**) *M. longifolium*; (**R**) *M. mimulifolium*; (**S**) *M. dicoelocarpum*; (**T**) *M. divaricatum*; (**U**) *M. tepicense*; (**V**) *M. repens*; (**W**) *M. montanum* var. *viridulum*; (**X**) *M. glabrum*; (**Y**) *M. nutans*; (**Z**) *M. perfoliatum*. Filled circles indicate satellites visible in most of the chromosomal spreads; open circles indicate satellites visible only in some spreads. (From Weiss-Schneeweiss et al., 2009: 1178)

evolutionary line could be documented with yet another chromosomal base number. Only two species exist in the section, *M. bibracteatum* and the very diminutive *M. repens*. Both have very small flowering heads and few florets, surrounded by only two phyllaries (hence the name of the section). Karyotypes for both species are unimodal, and the chromosomes are larger in *M. bibracteatum*, being the largest among all species in the genus (Fig. 5.8O). *M. repens* is diploid ($2n = 28$; Fig. 5.9V) and *M. bibracteatum* tetraploid ($2n = 56$). The karyotypes of these species are similar (except ploidy level), with one NOR found in *M. repens* and two in *M. bibracteatum.*

Sect. *Rhizomaria* contains only two species, *M. montanum* (with two varieties) and *M. aureum*. The former is uniformly diploid with $2n = 22$ (Fig. 5.7W), and the latter is hexaploid with $2n = 66$ (Fig. 5.8P). The morphology of the two species allows them to be easily identified, but most of the obvious features are quantitative, presumably reflecting the effects of polyploidization. It is interesting that in this complex of closely related perennial rhizomatous species only diploid and hexaploid plants are known, whereas one might anticipate also finding tetraploids; so far none has been detected. Due to the close morphological resemblance between the two species, and given lack of documented tetraploids, an autopolyploid origin for the hexaploid would seem most probable.

Sect. *Alcina* has been a mixed group of three species that, despite all having the same chromosome number of $2n = 22$ (Fig. 5.7X, Z), have been shown not to form an evolutionary group and now have been placed in three different sections (Stuessy et al., 2011). *M. perfoliatum* and *M. glabrum* have some morphological similarity, possessing broadly ovate outer phyllaries and dark brown-black, mostly smooth fruits at maturity that remain on the head in a crown after abscission of the sterile central disc florets (Fig. 3.2). The third species, *M. nutans*, despite having the same chromosome number of $2n = 22$ has now been removed into its own section *Nutantia* (Stuessy et al., 2011). Of the three species, *M. perfoliatum* has the largest chromosomes (Fig. 5.9Z).

Chapter 6
Reconstruction of Phylogeny

To fully understand systematic relationships and to infer evolutionary processes within a genus, it is essential to have a structure of phylogenetic affinities. Reconstruction of phylogeny is a hypothesis of macroevolution, involving longer-term evolutionary change, which has direct relevance for constructing classifications. Phylogenetics is crucial for making evolutionary inferences, especially in the identification of convergences and parallelisms as well as interpreting modes of speciation, the latter being particularly significant in deciphering allopolyploid evolutionary origins. In *Melampodium*, with so many different chromosome numbers and ploidy levels, critical phylogenetic analysis provides a foundation for revealing modes of chromosomal change.

Morphological Phylogenetic Analyses

Outside of isolated comments on miscellaneous phylogenetic relationships in *Melampodium* by Turner and King (1962), Stuessy (1971a), and Keil and Stuessy (1977), the first comprehensive analysis of the entire genus was done by Stuessy (1979). All known species of the genus at that time (37) were analyzed morphologically with 10 qualitative and 8 quantitative characters. These were selected based on experience with stability of features, plus the ability to assign evolutionary directionality (polarity) to the states, using stated assumptions. *Acanthospermum* was selected as an outgroup taxon. The method of tree building was a combination of Wagner Groundplan/Divergence (Wagner, 1961) plus that of Whiffin and Bierner (1972). This involved first establishing levels of evolutionary divergence based on total numbers of derived character states, with connections among taxa between levels based on the smallest number of evolutionary differences (Manhattan Distance).

T. F. Stuessy et al., *Systematics, Ecology, and Chromosomal Evolution of* Melampodium *(Asteraceae)*, https://doi.org/10.1007/978-3-031-81995-7_6

The results of the cladistic analysis based on a combination of qualitative and quantitative characters were presented as a network, rooted by a hypothetical ancestor (Fig. 6.1). This diagram reveals connections among the taxa of *Melampodium*, with bars indicating derived character state changes. The circles are the species and ploidy level is indicated. The general habitat in which each species typically occurs is also shown by the different backgrounds (tropical forest, pine-oak forest, and desert; these data were not used in the analysis). Section *Melampodium* ($x = 10$) and Sect. *Zarabellia* ($x = 9$) are the closest to the ancestral condition. Sects. *Bibractiaria* and *Rhizomaria* are shown as independently derived from Sect. *Melampodium*. Section *Alcina* appears to have had three independent origins from Sect. *Zarabellia*, and Sect. *Serratura* has two, one from Sect. *Zarabellia* and another from Sect. *Alcina*. The most primitive species are found in pine-oak forests, which were suggested as the ancestral habitat for the genus.

Separate analyses were also done in *Melampodium* using only qualitative or quantitative morphological characters. The results showed poor resolution with

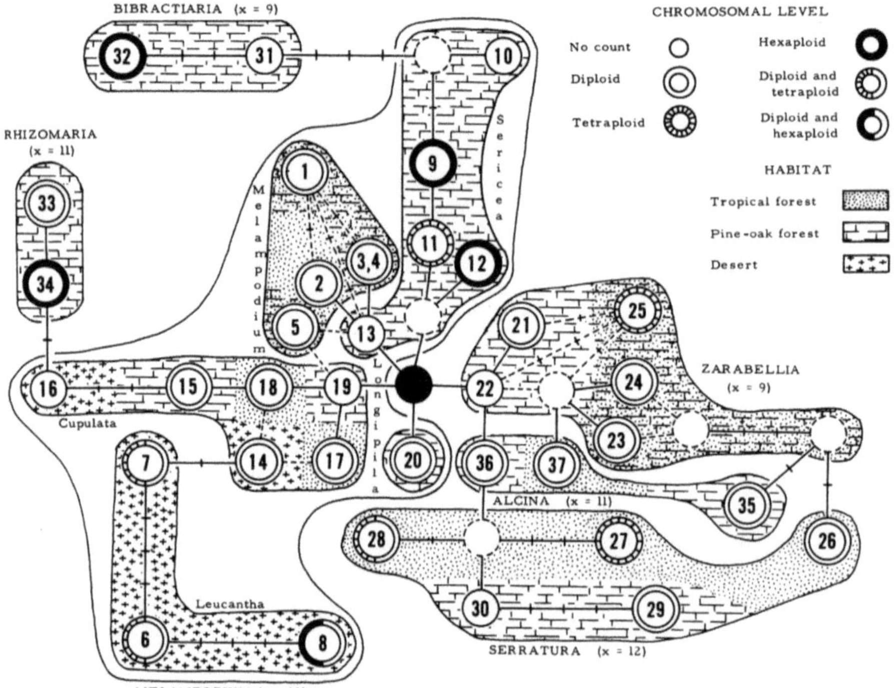

Fig. 6.1 Cladogram of 37 species of *Melampodium* (numbers in circles following Stuessy, 1972) based on selected qualitative and quantitative morphological characters. Dashed circles without numbers refer to hypothetical taxa; dashed lines connecting taxa refer to equally probable cladistic pathways. The black circle represents the ancestral taxon of the genus (with all ancestral conditions of character states). Length of lines (with bar markers) connecting taxa reveals the number of evolutionary character state changes. Chromosomal level and habitat occurrence of species are shown in the key. (From Stuessy, 1979: 183)

only the quantitative features, but better resolution with qualitative characters (10 used; Fig. 6.2). Due to the low number of characters used, resolution among series within Sect. *Melampodium* was not feasible, because several species retained the same set of characters. Nonetheless, some inferences could be made. Sect. *Melampodium* continued to be inferred as ancestral, whereas Sect. *Zarabellia* was derived from the former. Sect. *Rhizomaria* diverges from *M. sinuatum* in Sect. *Melampodium* (as with combined data), but Sect. *Bibractiaria* now derives from

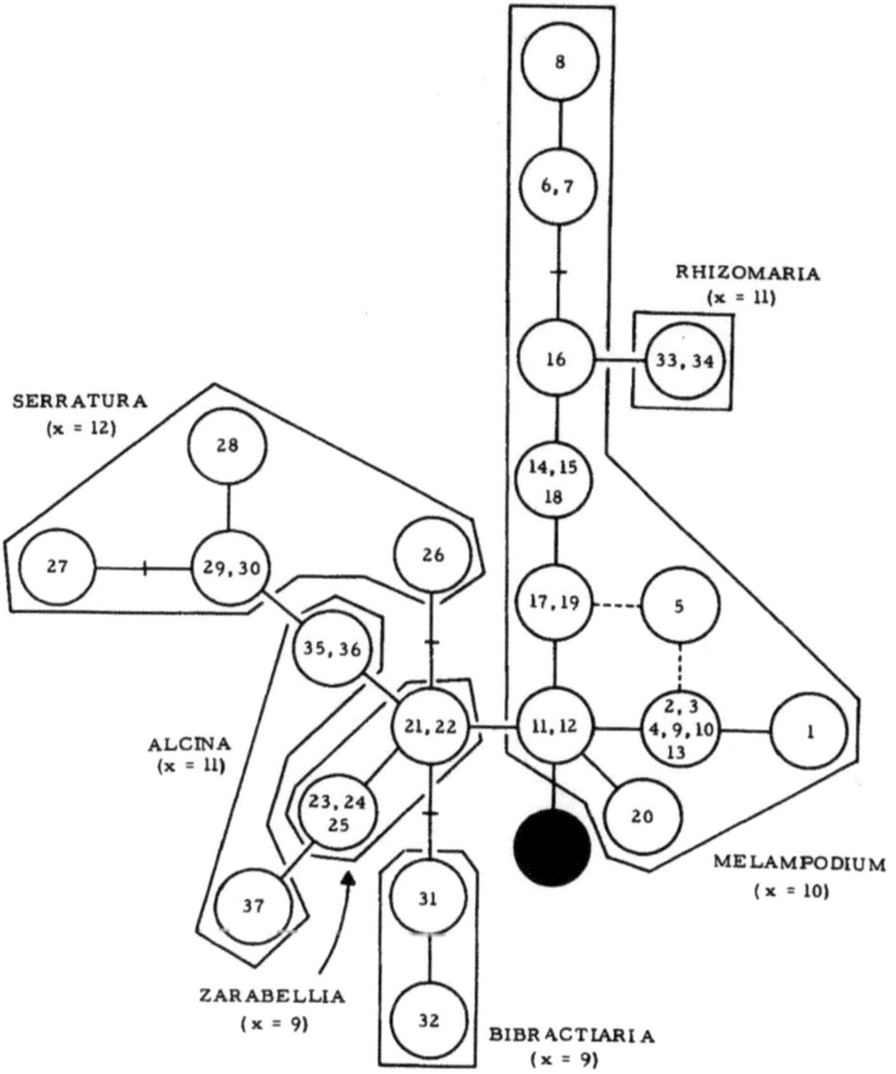

Fig. 6.2 Cladogram of 37 species of *Melampodium* based upon ten qualitative morphological characters. Method of data display similar to Fig. 6.1. (From Stuessy, 1979: 185)

Sect. *Zarabellia* rather than from Sect. *Melampodium*. Sects. *Serratura* and *Alcina* are shown as biphyletic.

These morphological cladistic analyses suggested several points regarding the phylogeny of *Melampodium*. Sect. *Melampodium*, based chromosomally on $x = 10$, was shown to be ancestral for the genus, and the pine-oak forest zone was inferred as the ancestral habitat. Although the outgroup genus, *Acanthospermum*, is known as $n = 11$, the more conspicuous ovary of the sterile disc florets (see Fig. 3.5; also in the outgroup) seemed important as a primitive reproductive condition, which supported Sect. *Melampodium* ($x = 10$) as ancestral. One of the most derived groups was the white-rayed complex (Ser. *Leucantha*; species 6–8 in Fig. 6.2) in Sect. *Melampodium*, and it derived from *M. sinuatum* (species 16) of Ser. *Cupulata*, which has a distribution at the tip of Baja California. This result contradicted the earlier hypothesis (Stuessy, 1971b) of origin from *M. americanum* (species no. 1 in Fig. 6.2) on the eastern side of the Sierra Madre Oriental (Fig. 9.6).

The most prominent result from the cladistic analyses is the frequency of multiple origins of sections and series. Particularly conspicuous are the tri- and biphyletic origins of Sects. *Alcina* and *Serratura*, respectively. Sect. *Alcina* consists of three species, *M. perfoliatum*, *M. glabrum*, and *M. nutans*, all known chromosomally as $x = 11$. At the time, it was realized that the latter species was morphologically distinct from the other two, but this had seemed to be the best placement initially (Stuessy, 1970a, 1972), obviously heavily influenced by the shared chromosome number.

Molecular Phylogenetic Analyses

The use of morphological data for phylogeny reconstruction using early cladistic algorithms was limited to some extent by the complexity of genetic control, evolutionary background, and adaptive change of morphological characters. From a historical perspective, after the structure of DNA was inferred by Watson and Crick (1953), it was quickly appreciated that this large macromolecule, composed of linear sequences of base pairs, could hold vast amounts of phylogenetic information useful for classification and understanding processes of evolution (e.g., Zuckerkandl & Pauling, 1965). Proteins held a similar potential, although more removed from the basic hereditary material, but the advantage was ease of isolation and amino acid sequencing, which preceded techniques for convenient DNA sequencing. Into the 1990s, plastid DNA was extracted, sheared, and these fragments compared among taxa (Palmer, 1987). From then on the methods progressed rapidly to sequencing of entire DNA fragments from the organelles and nucleus. It soon became possible to obtain huge quantities of DNA data with moderate financial investment, ushering in the age of genomics, bioinformatics, and many other molecular investigations (Hillis et al., 1996; Soltis et al., 1998). It was appropriate, therefore, to examine the phylogeny of *Melampodium* by using newer molecular methods. The previous morphological analyses had indicated some good units in classification within the genus, but some sections were questionable (e.g., Sect. *Alcina*).

Entire Genus

One of the first phylogenetic objectives was to confirm the proper placement of *Melampodium* in relation to other genera of the Heliantheae alliance. A pertinent study that contained five species of *Melampodium*, two of *Acanthospermum*, and one of *Lecocarpus*, plus numerous other genera in and around the páramo genus *Espeletia*, was that of Rauscher (2002). Nuclear ITS sequences were used, as these had been shown efficacious at the intra- and intergeneric level in Compositae (Baldwin et al., 1995). The results (Fig. 6.3) helped provide a perspective on where *Melampodium* fits among these other genera.

The phylogram from Rauscher (Fig. 6.3) shows *Melampodium*, *Acanthospermum*, and *Lecocarpus* distinct from all other sampled genera. The coverage of taxa is broad, including not only different parts of the *Espeletia* complex, but also *Smallanthus* (formerly *Polymnia*), *Milleria*, *Sigesbeckia*, and many others. It is possible that some genera that might be more closely related to *Melampodium* and its close allies were not included, but this seems doubtful based on past opinions regarding morphological affinities. Regardless, the *Melampodium* group was recovered as distinct from the rest of the species/genera. Hence, for selection of outgroups for new phylogenetic analyses, any of these other genera outside the *Melampodium* group might suffice. More importantly, *Melampodium* is seen as being close to *Acanthospermum* and *Lecocarpus*, cementing the morphological unity that has been emphasized for a long time. *Melampodium* is also a clear monophyletic unit as is also the complex of *Lecocarpus* and *Acanthospermum*. *Lecocarpus* may be interpreted as evolving from out of *Acanthospermum*, which is extremely likely as a long-distance dispersal event from the Ecuadorean mainland to the Galápagos Islands and subsequent in situ speciation. Within *Melampodium*, even if represented only by five species, some structure can be seen. *M. americanum* and *M. pilosum*, closely related in Sect. *Melampodium* Ser. *Melampodium* join closely together, as do *M. paniculatum* and *M. gracile* in Sect. *Zarabellia*, and *M. divaricatum* stands alone in Sect. *Serratura*. These results provided very positive indications that a denser sampling within *Melampodium* should reveal phylogenetic relationships and aid classification.

The first comprehensive molecular phylogenetic analysis of 115 samples involving all 40 species of *Melampodium* and selected outgroups was completed by Blöch et al. (2009) using nuclear ITS sequences and plastid *mat*K. The latter had been shown efficacious at the infrageneric level, especially in Compositae (e.g., Samuel et al., 2006). The ITS samples included all species of *Melampodium* (2–6 accessions for some species), three species of *Acanthospermum*, two of *Lecocarpus*, and one species each of *Galinsoga*, *Milleria*, *Sigesbeckia*, *Smallanthus*, and

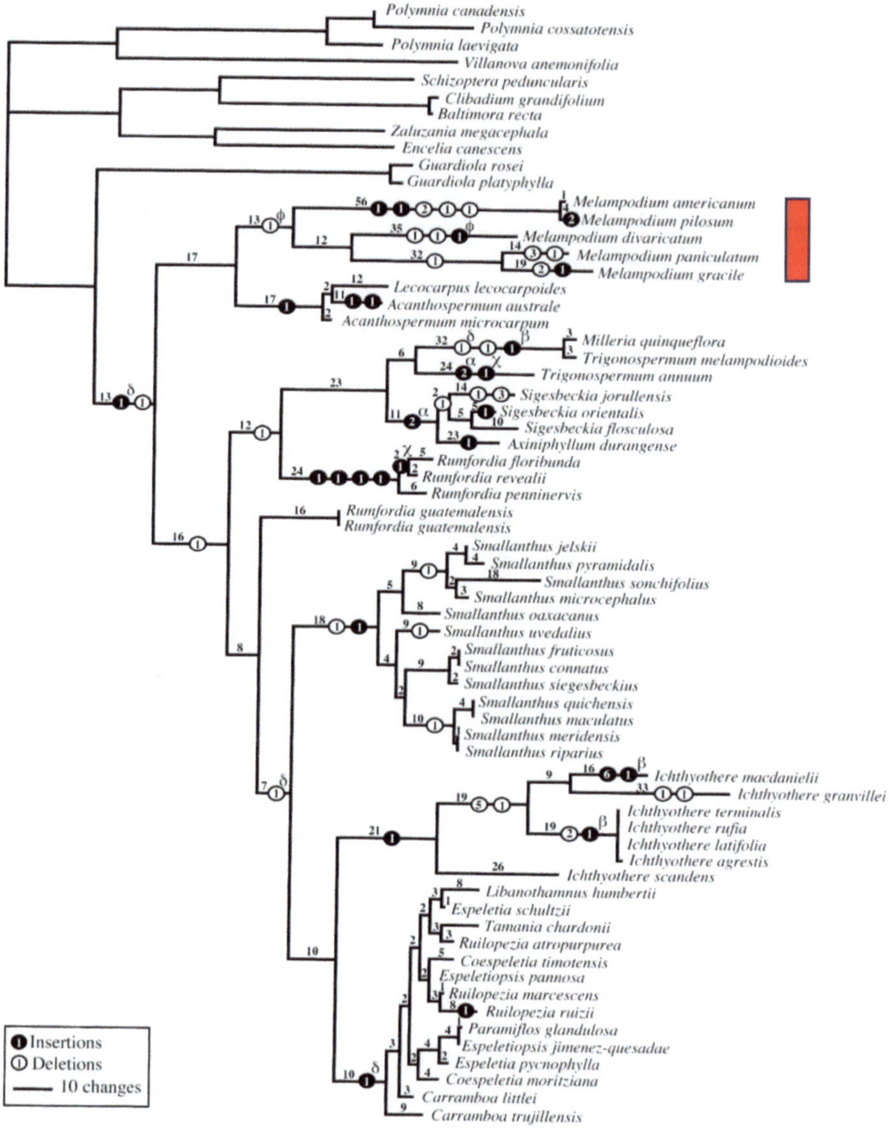

Fig. 6.3 Maximum parsimony phylogram based on ITS sequences of genera in the Heliantheae alliance related to *Melampodium*. Numbers above branches are branch lengths; insertions (black circles) and deletions (white circles) are shown on branches, with numbers indicating nucleotide length of each; symbols indicate homoplasious changes. From Rauscher (2002: 1078). Red bar highlights species of *Melampodium*

Trigonospermum. Results from the nr ITS region with both maximum parsimony and Bayesian analyses revealed a highly resolved phylogenetic pattern (Fig. 6.4). Of initial interest is that the closely related genera, *Acanthospermum* and *Lecocarpus*,

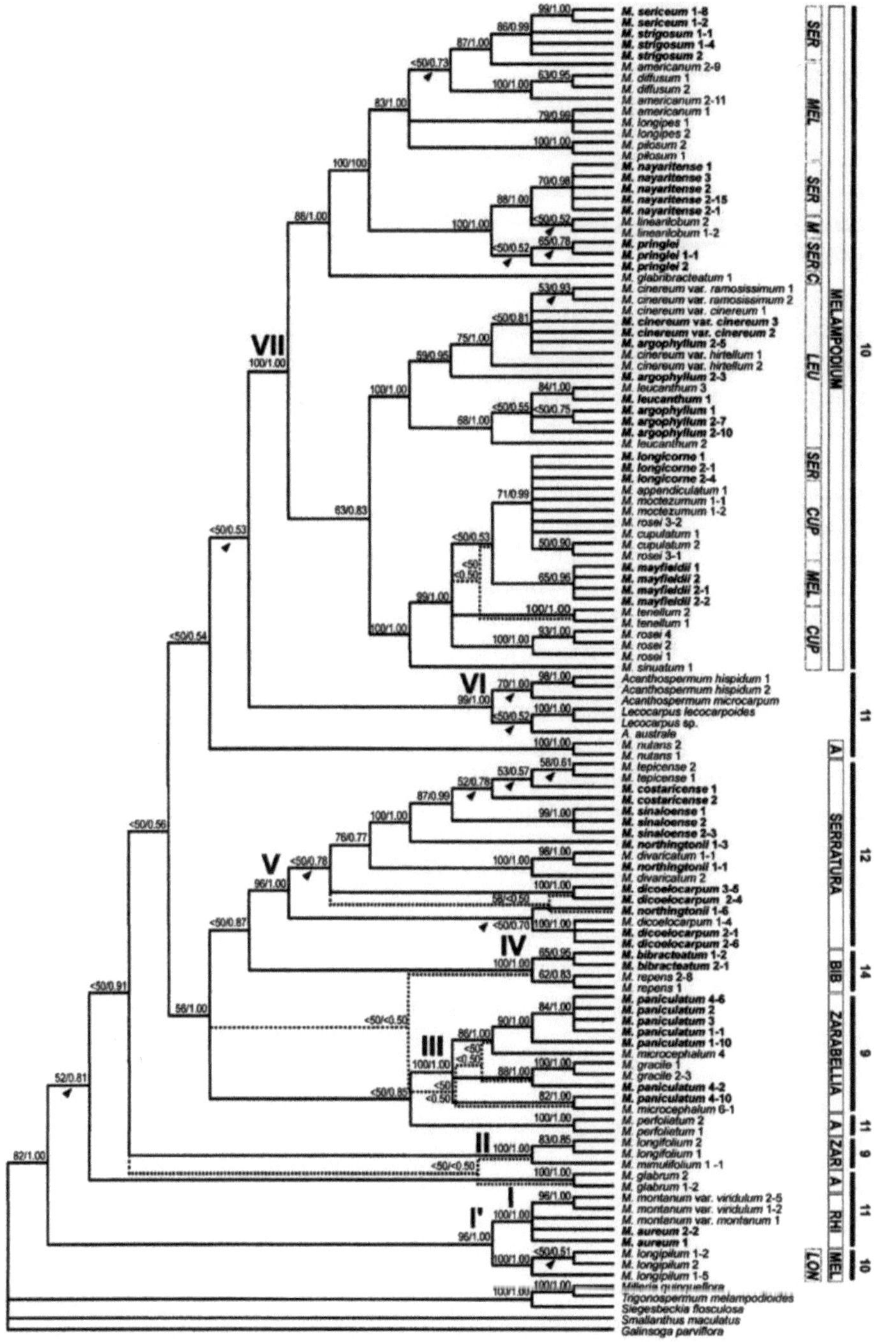

Fig. 6.4 Phylogenetic relationships of species of *Melampodium* and related genera inferred from Bayesian (solid lines) and maximum parsimony analysis (dotted lines) of the ITS region. Branches collapsing in the strict consensus tree are indicated by arrowheads. Numbers at nodes are bootstrap values/posterior probabilities. Roman numerals refer to major evolutionary clades. Numbers after species names indicate different accessions and clones (after hyphens). Polyploid taxa are shown in bold. Vertical bars on the right show the taxonomic classification of the genus (after Stuessy, 1972) with basic chromosome numbers. Sections: *A* Sect. *Alcina*, *BIB* Sect. *Bibractiaria*, *RHI* Sect. *Rhizomaria*, *ZAR* Sect. *Zarabellia*. Series in Sect. *Melampodium*: *CUP* Ser. *Cupulata*, *LEU* Ser. *Leucantha*, *LON* Ser. *Longipilum*, *MEL and M* Ser. *Melampodium*, *SER* Ser. *Sericea*. (From Blöch et al., 2009: 225)

form together with *M. nutans* a group that is sister to Sect. *Melampodium*. *Lecocarpus* appears to originate from within *Acanthospermum*, as suggested earlier.

Among Sections and Series

Molecular phylogenetic analyses also reveal a correlation of clade structure with the taxonomic sections and series proposed earlier on morphological and cytological grounds (Stuessy, 1972). With ITS (Fig. 6.4), Sect. *Melampodium* (clade VII) is distinct from the other sections of the genus and with high bootstrap support (100), except for *M. longipilum*, which instead forms with Sect. *Rhizomaria* a clade that is sister to the remainder of *Melampodium*. Relationships among series in Sect. *Melampodium* are also instructive. Unsurprisingly, the white-rayed complex (Ser. *Leucantha*) is very well supported (100 BS). *M. cinereum* and *M. leucanthum* are separated clearly, but the position of the hexaploid *M. argophyllum* is ambiguous, with some accessions joining with those of the other two species, perhaps suggesting allopolyploid origin. Ser. *Cupulata* contains those species recognized by Stuessy (1972), except for the additional inclusion of *M. mayfieldii*, regarded initially as belonging to Ser. *Melampodium* (perhaps close to *M. americanum*) when it was originally described by Turner (1993). *M. longicorne* also is recovered in this series, previously placed in Ser. *Sericea* (Stuessy, 1972). Three other sections, *Serratura*, *Bibractiaria*, and *Zarabellia* are distinct (clades III, IV, V), except that *M. longifolium* and *M. mimulifolium* lie outside Sect. *Zarabellia*. It is noteworthy that some accessions of the tetraploid species, *M. paniculatum*, of suspected allopolyploid origin (Stuessy & Brunken, 1979), tie to the diploid *M. gracile* and *M. microcephalum*. Sect. *Rhizomaria* is distinct, but *M. longipilum*, placed by Stuessy (1972) in its own series in Sect. *Melampodium*, joins with it. The species in Sect. *Alcina* that were problematic in the morphological cladistic analyses (Stuessy, 1979), i.e., *M. nutans*, *M. glabrum*, and *M. perfoliatum*, occur widely separated in different clades.

Results from *mat*K (Fig. 6.5; Blöch et al., 2009) provide interesting contrasts with the data from ITS. Most of the sections of the genus are recovered in the analysis, i.e., Sects. *Melampodium*, *Serratura*, *Bibractiaria*, *Zarabellia*, and *Rhizomaria*. In fact, the sectional structure is clearer than with ITS. Together with *M. glabrum*, *Acanthospermum* and *Lecocarpus* are sister to the remainder of *Melampodium*. Sect. *Alcina* is polyphyletic.

Data from ITS and *mat*K were combined into a consensus network (Holland et al., 2004), which reveals all splits between species within a set of trees (Fig. 6.6; Blöch et al., 2009), with the ellipses indicating the disposition of species following the classification of Stuessy (1972). The large Sect. *Melampodium* is well removed from the other sections, and Sects. *Rhizomaria*, *Zarabellia*, *Serratura*, and *Bibractiaria* are distinct from the others on the network. Unsurprisingly, the three species of Sect. *Alcina*, *M. perfoliatum*, *M. glabrum*, and *M. nutans*, fall in different places and underline the non-monophyletic nature of the group. *Melampodium*

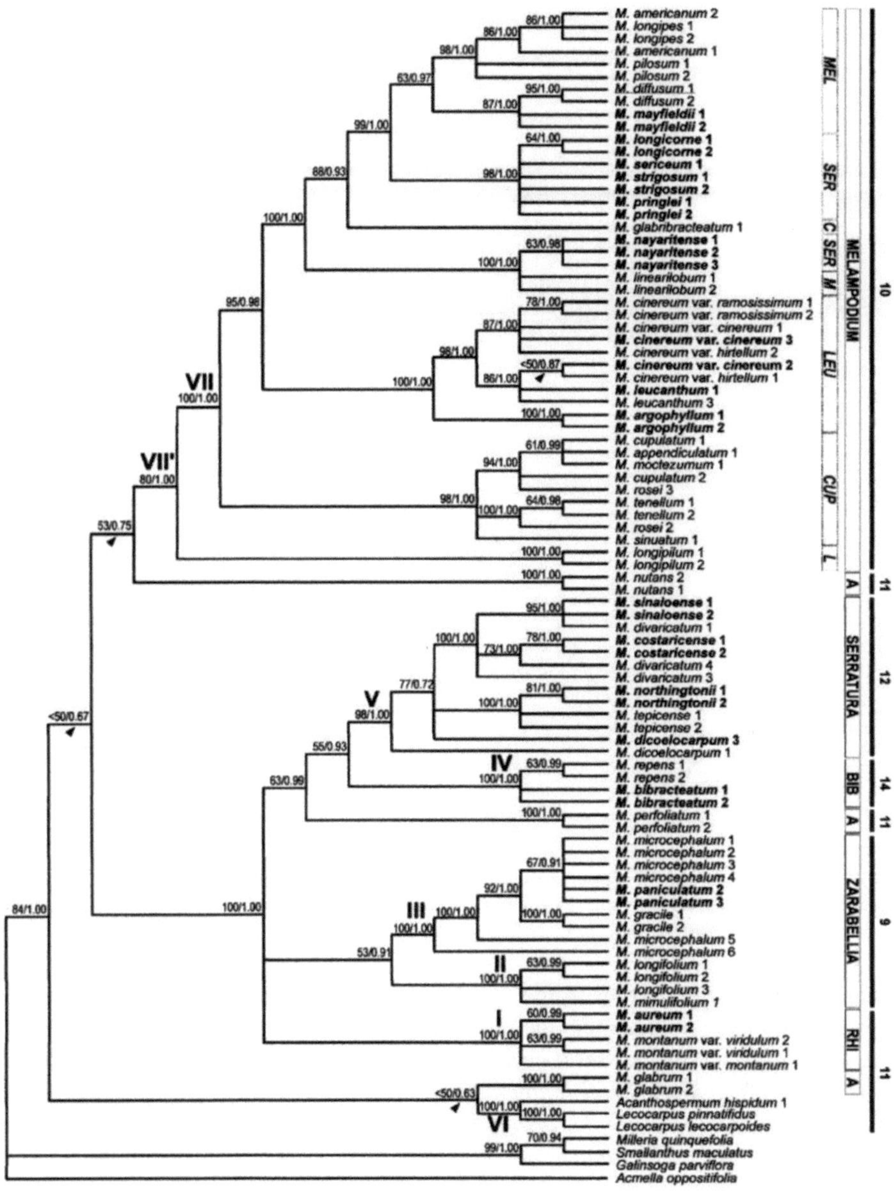

Fig. 6.5 Phylogenetic relationships of *Melampodium* and related genera inferred from Bayesian analysis of the plastid *mat*K gene. Branches that collapse in the strict consensus tree are indicated by arrowheads. Numbers at nodes are bootstrap values/posterior probabilities. Numbers after species names refer to different accessions and clones (after dashes). Polyploid taxa are indicated in bold. Major evolutionary clades are shown by Roman numerals. Basic chromosome numbers and classification of the genus (after Stuessy, 1972) are shown vertically on the right side. Abbreviations as in Fig. 6.4. (From Blöch et al., 2009: 227)

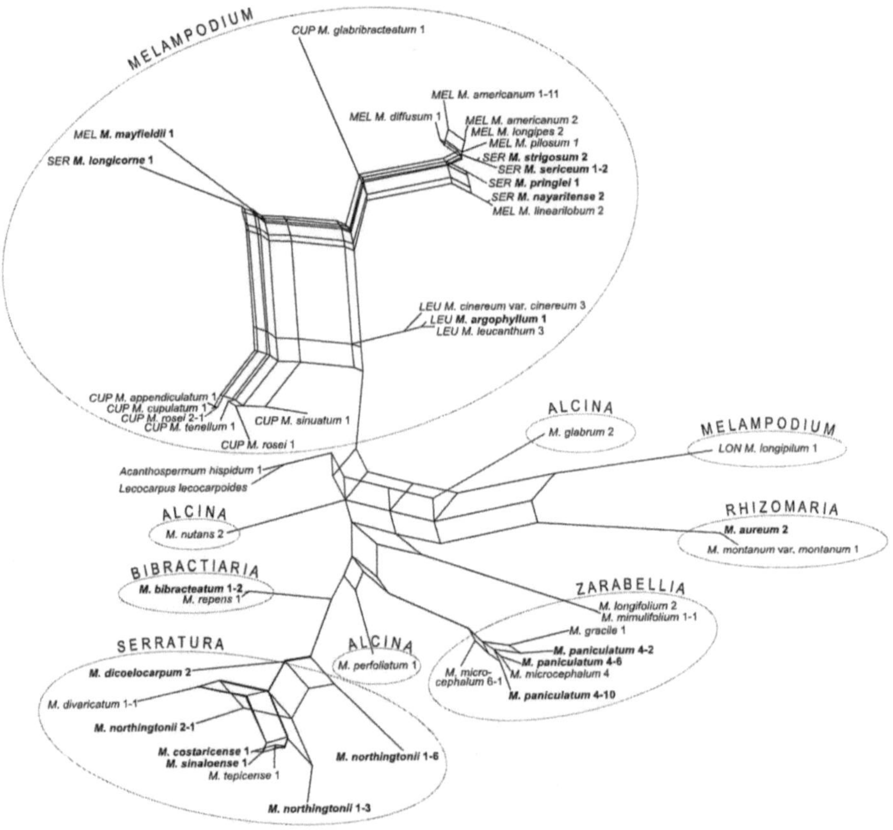

Fig. 6.6 Consensus network among species of *Melampodium* from 50 trees of the sets of posterior trees from ITS and *mat*K data sets. Polyploid taxa are indicated in bold. Three-letter prefixes refer to series within Sect. *Melampodium* (see legend of Fig. 6.4). From Blöch et al. (2009: 228)

longipilum is also an outlier from Sect. *Melampodium*, reflecting its different position in plastid and nuclear genome phylogenetic trees. Within Sect. *Melampodium*, the white-rayed complex (Ser. *Leucantha*) is well removed from the other clusters, as is also Ser. *Cupulata*. Ser. *Melampodium* forms a unit, except that the polyploids, *M. strigosum*, *M. sericeum*, *M. pringlei,* and *M. nayaritense*, from Ser. *Sericea*, are embedded within it. *Melampodium glabribracteatum* is very isolated as are the polyploids *M. mayfieldii* and *M. longicorne*, the latter two being of interserial allopolyploid origin.

Additional molecular phylogenetic studies have been performed on all species of the large Sect. *Melampodium* (Stuessy et al., 2011). This is of special interest because the section contains 22 species, more than half of the entire genus, and with many polyploid species (7 out of 12 total, not counting the infraspecific polyploid races in *M. dicoelocarpum*, *M. cinereum*, and *M. leucanthum*). Analyses were carried out with 5S rDNA non-transcribed spacer (Fig. 6.7), low-copy nuclear *PgiC*I

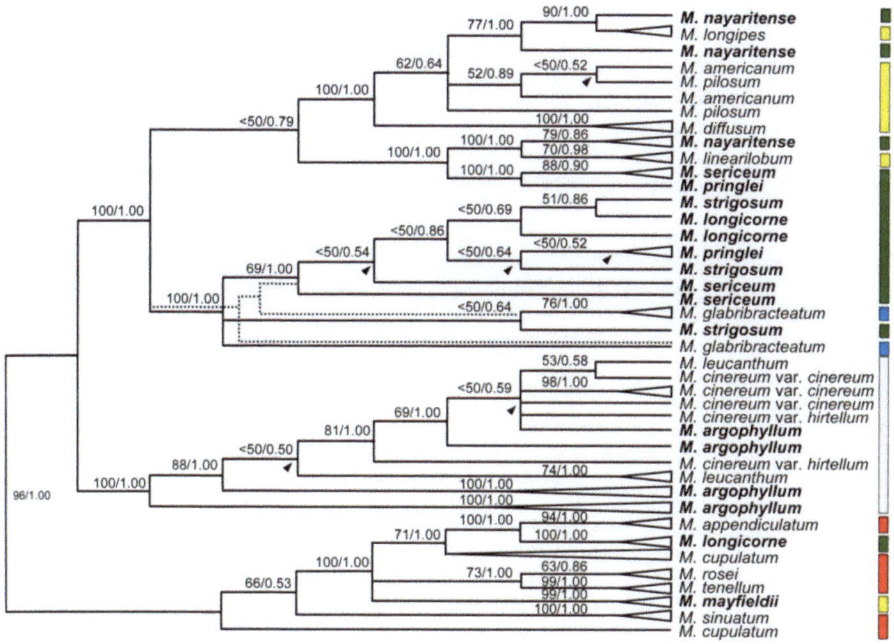

Fig. 6.7 Phylogenetic relationships among species of section *Melampodium* based on nuclear 5S rDNA non-transcribed spacer region inferred from Bayesian (solid lines) and maximum parsimony analysis (dotted lines). Series shown by colors: red, *Cupulata*; blue, *Glabribracteata*; gray, *Leucantha*; yellow, *Melampodium*; green, *Sericea*. Polyploid taxa indicated in bold. Branches that collapsed in maximum parsimony strict consensus are indicated by arrowheads. (From Stuessy et al., 2011: 443)

and *Pgi*CII (Fig. 6.8), and combined plastid *mat*K and *psb*A-*trn*H regions (Fig. 6.9). Five series had been recognized previously (Stuessy, 1972) in this section: *Melampodium*, *Leucantha*, *Sericea*, *Cupulata*, and *Longipila*. In all analyses the white-rayed complex, Ser. *Leucantha*, remains distinct as a good unit, as does Ser. *Cupulata*. Ser. *Longipila*, with the single species, *M. longipilum*, continues to be divergent, and it seems best to leave it as a monospecific unit. *M. glabribracteatum* had been placed in Ser. *Cupulata*, based primarily on the scarious margins of the outer phyllaries, but with molecular data it is quite distinct, better placed in its own monospecific series.

The most interesting molecular relationships occur between Ser. *Melampodium* and Ser. *Sericea*. Cytogenetic studies (see Chap. 7) have revealed that *M. linearilobum* in Ser. *Melampodium* has been one of the progenitors in the allopolyploid origins of *M. nayaritense* (4*x*), *M. sericeum* (6*x*), and *M. pringlei* (6*x*); *M. strigosum* (4*x*) is the other parent for the origin of the two listed hexaploids. *Melampodium mayfieldii* (4*x*) and *M. longicorne* (6*x*) are interserial hybrids between parents in Ser. *Cupulata* and Ser. *Melampodium*. The allopolyploid origin of the tetraploids and hexaploids suggests leaving these in Ser. *Sericea* and keeping the diploids in Ser.

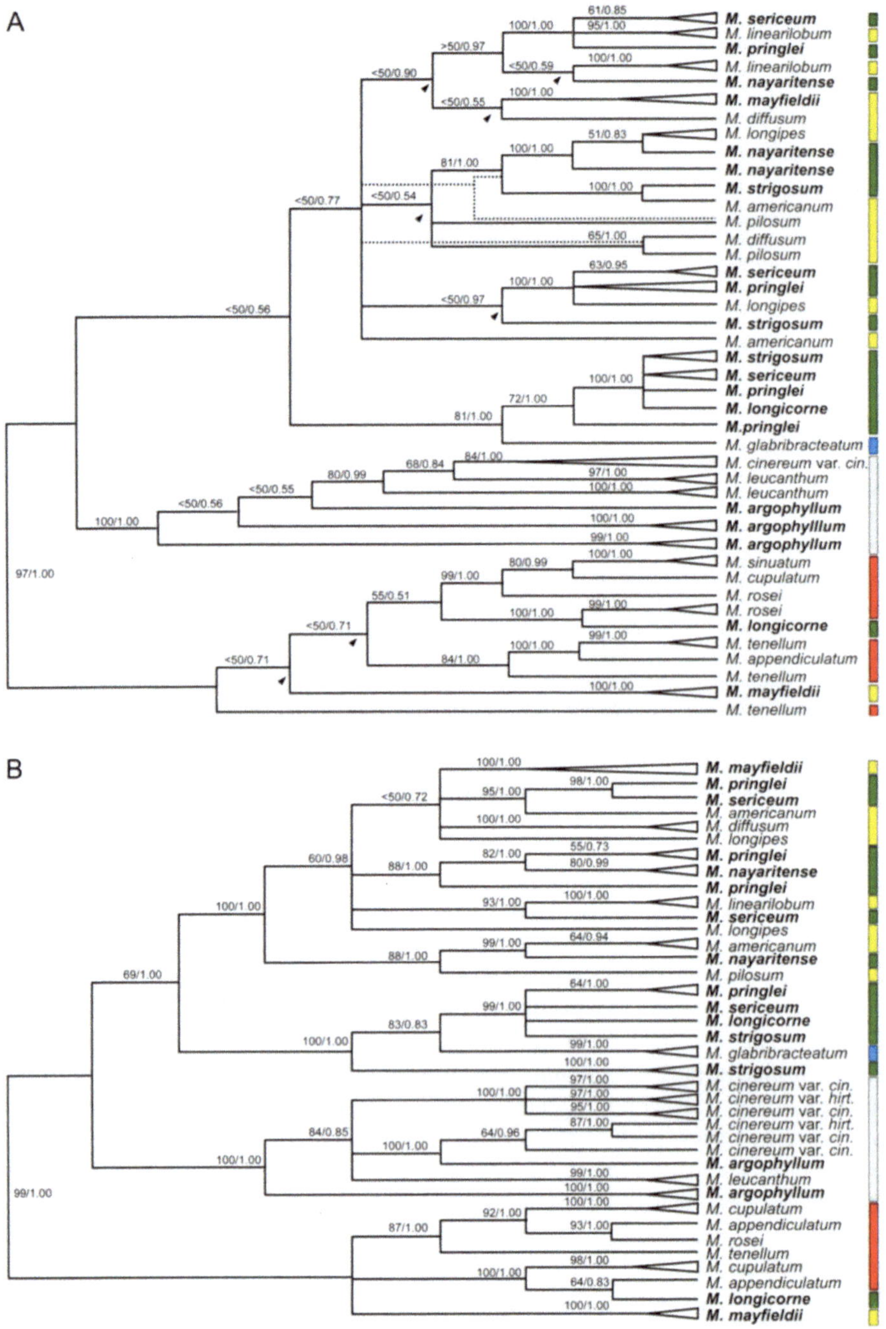

Fig. 6.8 Phylogenetic relationships among species of Sect. *Melampodium* (excluding the divergent *M. longipilum*) based on (**A**) low-copy nuclear *Pgi*CI and (**B**) *Pgi*CII sequences. Series indicated by colors: red, *Cupulata*; blue, *Glabribracteata*; gray, *Leucantha*; yellow, *Melampodium*; green, *Sericea*. Polyploid taxa are indicated in bold. Branches collapsing in the maximum parsimony strict consensus trees are indicated by arrowheads. (From Stuessy et al., 2011: 444)

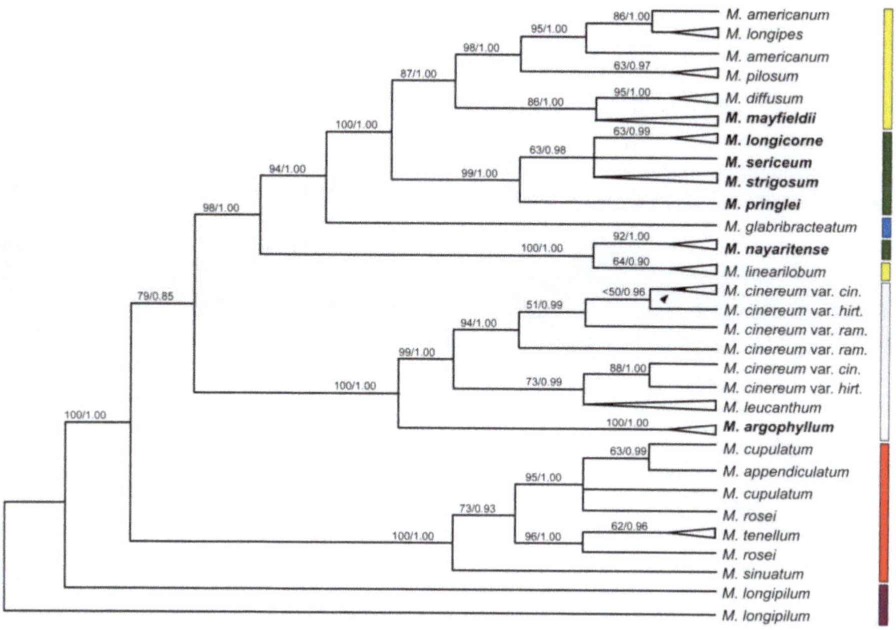

Fig. 6.9 Summary of phylogenetic relationships among species of Sect. *Melampodium* based on analysis of combined plastid *mat*K and *psb*A-*trn*H data sets. Series indicated by colors: red, *Cupulata*; blue, *Glabribracteata*; gray, *Leucantha*; purple, *Longipila*; yellow, *Melampodium*; green, *Sericea.* Polyploid taxa are indicated in bold. Branches collapsing in the maximum parsimony strict consensus tree are indicated by arrowheads. (From Stuessy et al., 2011: 445)

Melampodium, except for the tetraploid *M. mayfieldii* also included in Ser. *Melampodium*. The plastid markers (Fig. 6.9) track more precisely the series limits than do the nuclear markers (Fig. 6.7 and 6.8), perhaps because they do not reflect as precisely the reticulate patterns.

Within the White-Rayed Complex

Populational-level molecular analyses were completed on the species of the white-rayed complex (Rebernig et al., 2012), of interest because the group is isolated geographically and distinct morphologically from the rest of the genus. Particular focus has been on the taxonomic placement and evolutionary origin of the hexaploid, *M. argophyllum*, in relation to the diploids *M. cinereum* and *M. leucanthum* (each also with autotetraploid races). A hypothesis was tested for an allopolyploid origin of *M. argophyllum* involving the two diploid relatives.

Because of the close relationships of taxa in the white-rayed complex, Rebernig et al. (2012) obtained resolution by using population markers: AFLPs, and three non-coding plastid regions, *psb*A-*trn*H, *rpl*32-*trn*L, and *ndh*F-*rpl*32. Collections

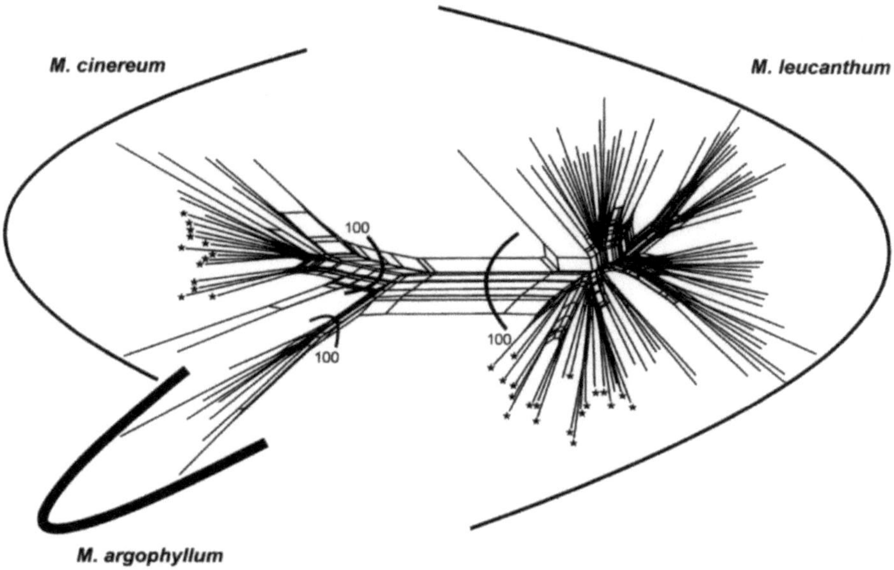

Fig. 6.10 Neighbor-net for the AFLP data using Nei-Li distances among species of the white-rayed complex of *Melampodium*. Numbers indicate bootstrap values from a neighbor joining analysis. Asterisks indicate tetraploid cytotypes. (From Rebernig et al., 2012: 1047)

were made from 90 populations of *M. leucanthum*, 22 populations of *M. cinereum*, and 2 of *M. argophyllum*. The results from the plastid data reveal some variation depending upon the marker, but in general, three distinct entities are recovered that correspond to the three taxonomic species (Blöch et al., 2009; Stuessy et al., 2011; Rebernig et al., 2012). With analyses of the AFLP data (Fig. 6.10), again three clusters of lineages (with Nei-Li distances) are seen, with *M. argophyllum* joining preferentially toward *M. cinereum*, but still distinct from it. Another method, restriction patterns from nrITS sequences (35S rRNA genes) was employed. Figure 6.11 gives evidence that the varieties of *M. cinereum* are all distinct from *M. leucanthum*, both in ITS1 and ITS2, especially so in the latter. Some of the populations of *M. argophyllum* appear to have inherited material from both *M. cinereum* and *M. leucanthum*, particularly demonstrated by ITS2. With ITS1, the hexaploid appears more closely related to *M. leucanthum* than to *M. cinereum*.

Although there is no present indication of hybridization between any of the three species, there is evidence of possible genetic exchange in the origin of the entire complex. The data suggest some introgression into *M. cinereum* var. *hirtellum* (western race) from *M. leucanthum*, as indicated by *leucanthum*-type plastid haplotypes and ITS sequences in some var. *hirtellum* populations. The data support that *M. argophyllum*, due to its genetic distinctness, may be of autopolyploid, rather than allopolyploid origin, perhaps representing a relic lineage from the ancestral taxon.

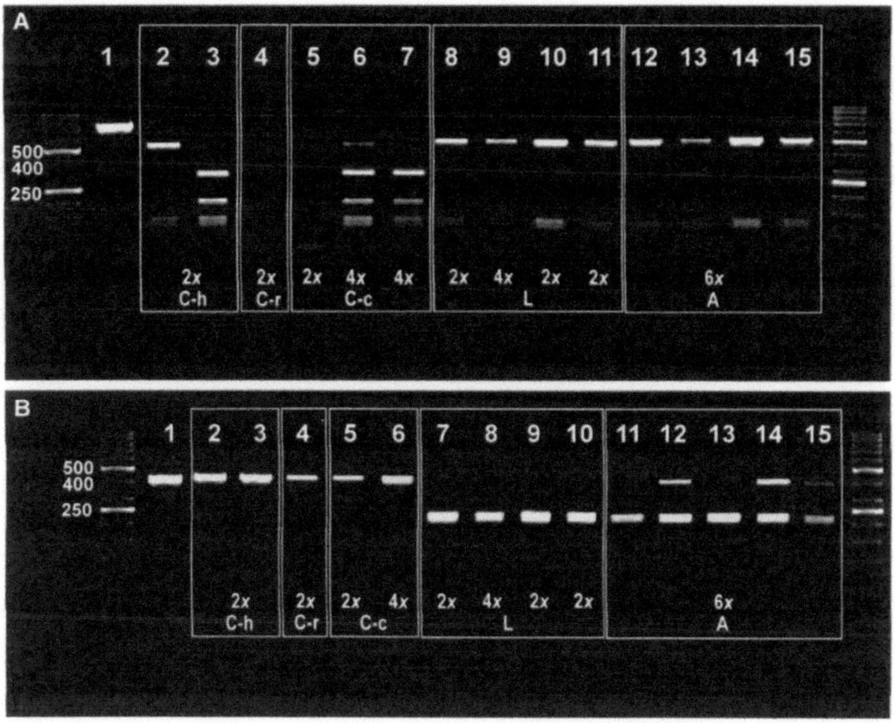

Fig. 6.11 Restriction patterns for ITS from the white-rayed species of *Melampodium*. (**A**) The entire ITS digested with *Bst*UI: *1*, undigested ITS1 amplification product; *2*, *M. cinereum* var. *hirtellum* 19057C, 2*x*; *3*, *M. cinereum var. hirtellum* 19061B, 2*x*; *4*, *M. cinereum* var. *ramosissimum* 19065H, 2*x*; *5*, *M. cinereum* var. *cinereum* 18688H, 2*x*; *6*, *M. cinereum* var. *cinereum* 18697F, 4*x*; *7*, *M. cinereum* var. *cinereum* 18694J, 4*x*; *8*, *M. leucanthum* 18814B, 2*x*; *9*, *M. leucanthum* 18785G, 4*x*; *10*, *M. leucanthum* 20033C, 2*x*; *11*, *M. leucanthum* 19056B, 2*x*; *12*, *M. argophyllum* 19059E, 6*x*; *13*, *M. argophyllum* 19060A, 6*x*; *14*, *M. argophyllum* 19060CI, 6*x*; *15*, *M. argophyllum* 19060E, 6*x*; (**B**) ITS2 only digested with *Hae*III: *1*, undigested ITS2 amplification product; *2*, *M. cinereum* var. *hirtellum* 19057B, 2*x*; *3*, *M. cinereum* var. *hirtellum* 19061B, 2*x*; *4*, *M. cinereum* var. *ramosissimum* 19065H, 2*x*; *5*, *M. cinereum* var. *cinereum* 18688H, 2*x*; *6*, *M. cinereum* var. *cinereum* 18694N, 4*x*; *7*, *M. leucanthum* 18814B, 2*x*; *8*, *M. leucanthum* 18785G, 4*x*; *9*, *M. leucanthum* 20033C, 2*x*; *10*, *M. leucanthum* 19056B, 2*x*; *11*, *M. argophyllum* 19059A, 6*x*; *12*, *M. argophyllum* 19059E, 6*x*; *13*, *M. argophyllum* 19059I, 6*x*; *14*, *M. argophyllum* 19060A, 6*x*; *15*, *M. argophyllum* 19060C, 6*x*. Abbreviations: C-h, *M. cinereum* var. *hirtellum*; C-r, *M. cinereum* var. *ramosissimum*; C-c, *M. cinereum* var. *cinereum*; L, *M. leucanthum*; A, *M. argophyllum*. (From Rebernig et al., 2012: 1049)

Nonetheless, an allopolyploid origin cannot be discounted completely, especially because the ITS data tend in this direction. More work on this hexaploid taxon needs to be done, especially at the populational level, but the species grows in habitats that are difficult to access.

Chapter 7
Chromosomal Evolution

Because *Melampodium* harbors such a wide range of different chromosome numbers, $n = 9, 10, 11, 12, 14, 18, 20, 23, 24, 27, 28, 30$, and 33, it is obvious that chromosomal change has been involved during speciation in the genus. The challenge is to examine these numbers in the context of new molecular phylogenetic information to obtain more robust hypotheses of chromosomal evolution. Because polyploidy has been involved in the origin of several of the chromosome numbers in the genus (tetraploids and hexaploids), it is important to distinguish between autopolyploids and allopolyploids and in the latter to determine the putative parents.

Early Hypotheses of Chromosomal Change

The first hypothesis on chromosomal evolution in *Melampodium* was proposed by Stuessy (1971a). Based on a revision of the genus (Stuessy, 1972), it was possible to seek correlations between these taxa and the different (base) chromosome numbers. With the numbers available at the time, four evolutionary groups and three ploidy levels were hypothesized (Fig. 7.1A). Four diploid chromosome numbers formed the basis for the polyploid series: $x = 9, 10, 11$, and 12 ($x = 14$ was unknown at that time; Weiss-Schneeweiss et al., 2009). Diploids, tetraploids, and hexaploids were found, based on both $x = 9$ ($n = 9, 18$, and 27), and $x = 10$ ($n = 10, 20$, and 30). Diploids and hexaploids were recovered within the $x = 11$ lineage ($n = 11, 33$, in the absence of tetraploids) and diploids and tetraploids within the $x = 12$ group ($n = 12$, 25 ± 1, and dysploidy on the tetraploid level with $n = 23$). The numbers given in parentheses in Fig. 7.1A refer to the number of species known at that ploidy level. The majority of karyologically known taxa had numbers based on $x = 10$, with 11 diploid, three tetraploid, and two hexaploid species known at that time (16 species total); the other chromosome base numbers ($x = 9, 11$, and 12) were found in only five species each. These groups correlated with taxonomic sections: Sect. *Zarabellia*,

T. F. Stuessy et al., *Systematics, Ecology, and Chromosomal Evolution of* Melampodium *(Asteraceae)*, https://doi.org/10.1007/978-3-031-81995-7_7

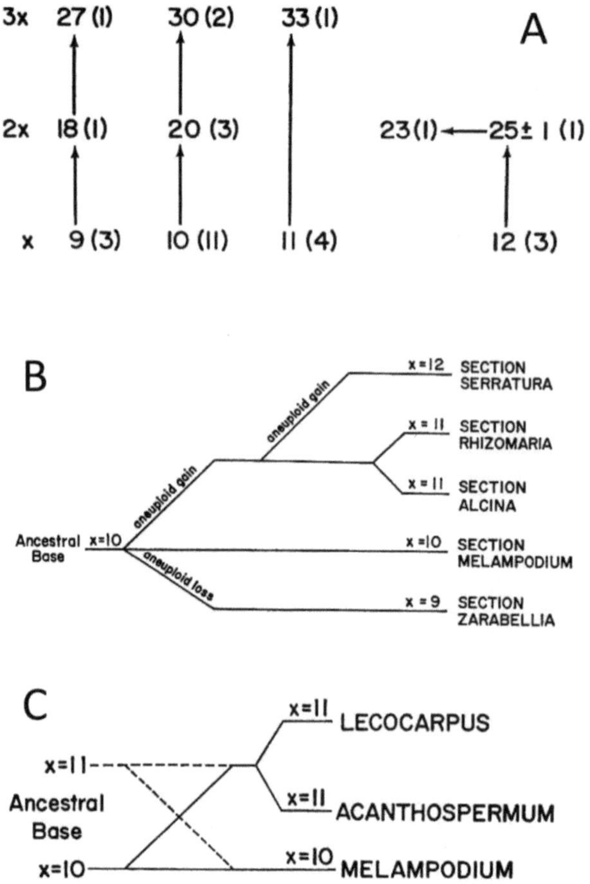

Fig. 7.1 Inferred chromosomal relationships among evolutionary lineages within *Melampodium* and among related genera. (**A**) Euploid relationships among known haploid chromosome numbers in *Melampodium*. The number of species counted at each chromosomal level is shown within parentheses. Species with polyploid races are recorded at two different levels. (**B**) Hypothetical evolutionary relationships among the four chromosomal lines within *Melampodium*. (**C**) Hypothetical evolutionary relationships among *Acanthospermum*, *Lecocarpus*, and *Melampodium*. Dashed lines suggest less probable evolutionary pathways. (From Stuessy, 1971a: 734)

$x = 9$; Sect. *Melampodium*, $x = 10$; and Sect. *Serratura*, $x = 12$, except for Sects. *Alcina* and *Rhizomaria* both based on $x = 11$.

The correlations of morphological and cytological data in *Melampodium* permitted a phylogenetic diagram to be constructed illustrating hypothesized evolutionary trends in chromosome number evolution (Fig. 7.1B). It was assumed that $x = 10$ was ancestral, because (1) this base number dominated among species in the genus (16/31 of species counted), and (2) species with $x = 10$ all had sterile ovaries of the disc florets with a conspicuous demarcation, which, considering that this condition is also known from the closely related genera *Acanthospermum* and *Lecocarpus*, was interpreted as plesiomorphic (ancestral) compared to only rudimentary ovaries in the other lineages (see Fig. 3.5). These considerations, therefore, suggested mechanisms of dysploid loss leading to $x = 9$, and two independent gains into $x = 11$ and 12. Figure 7.1C indicates that from an ancestral base chromosome number of $x = 10$, ascending dysploidy might have given rise to the $x = 11$ lineages of *Acanthospermum* and *Lecocarpus*. Alternatively, it was shown that $x = 11$ could have been ancestral for the complex with descending dysploidy leading to *Melampodium*. The available data did not provide evidence to reject either hypothesis definitively.

Testing the Hypotheses of Chromosomal Change

The first attempts to revisit the hypotheses of chromosomal evolution within *Melampodium* were by Blöch et al. (2009) and Weiss-Schneeweiss et al. (2009). Phylogenetic relationships within the genus inferred from analyses of nuclear ITS and plastid *mat*K regions provided a backbone for analyses of base chromosome number evolution (Fig. 7.2). In this analysis all 40 species of the genus were included, all with known chromosome numbers (Weiss-Schneeweiss et al., 2009). In some cases, numerous populations had been sampled, but in a few rare taxa (e.g., *M. mimulifolium*), the chromosome numbers were established for only one population. The cytological data mapped onto a phylogenetic network demonstrated that the same base chromosome numbers ($x = 11$) are shared by species within different sections and also by *Acanthospermum* and *Lecocarpus*, hence judged to be ancestral. The majority of species were based on $x = 10$, all in monophyletic Sect. *Melampodium*, except for *M. longipilum* (suggested as perhaps having had a different origin). This species has long been regarded as atypical in the section, treated by Stuessy (1972) as belonging in a distinct Ser. *Longipila*. The descending dysploidy series based on $x = 9$ gave rise to Sect. *Zarabellia*, with two distinct origins, being treated taxonomically as Ser. *Zarabellia* (with *M. mimulifolium* and *M. longifolium*) and Ser. *Tribracteata* (containing *M. paniculatum*, *M. microcephalum*, and *M. gracile*). The more chromosomally divergent lineages with $x = 12$ and 14 were hypothesized to have developed via ascending dysploidy to yield Sects. *Serratura* and *Bibractiaria*, respectively. Within each lineage, polyploidy took place to yield the chromosomal diversity observed.

Within the large Sect. *Melampodium*, subgroups representing taxonomic series could be identified. Unsurprisingly, the white-rayed complex, Ser. *Leucantha*, was very distinct in the analysis. Ser. *Cupulata* was also unified, except for *M. glabribracteatum*, formerly of the same series (Stuessy, 1972), but because of those molecular results, now placed in its own series (Ser. *Glabribracteata*; Stuessy et al., 2011). Ser. *Melampodium* formed a very good cluster, but species from Ser. *Sericea* were not recovered in one clade. These latter species are all either allotetraploid or allohexaploid, and they share ancestry with different sets of parental diploids mostly from Ser. *Melampodium*, and a few from Ser. *Cupulata*, which helps explain their integrated positions into these two series.

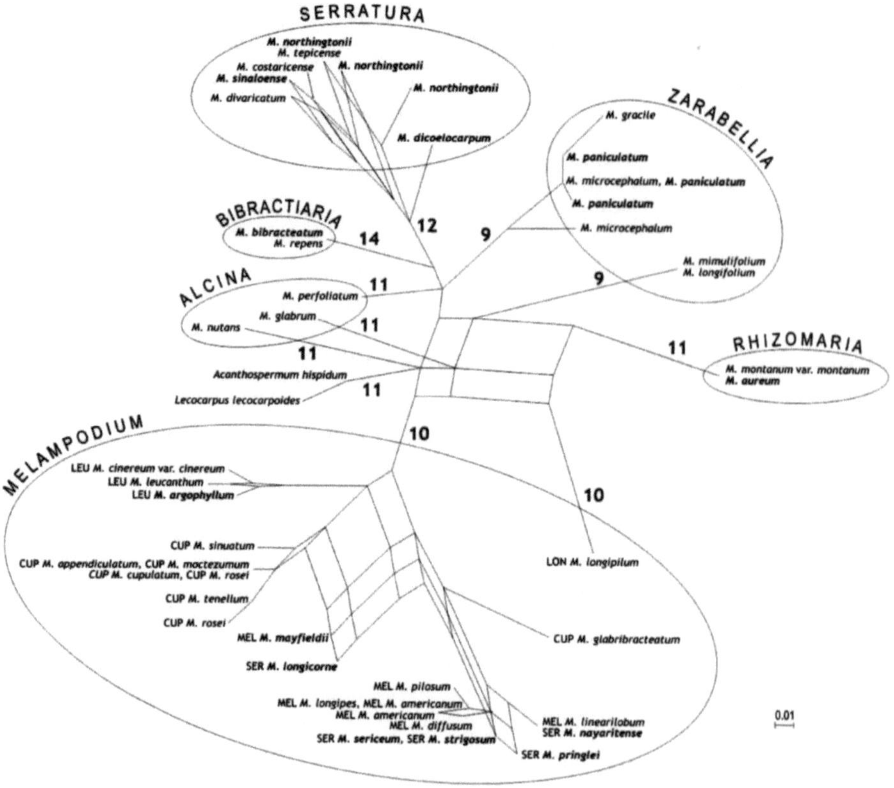

Fig. 7.2 Distribution of chromosome base numbers among lineages of *Melampodium* and related genera, on a consensus network from 50 Bayesian posterior trees from the nuclear ITS and the plastid *mat*K data set of Blöch et al. (2009). To aid legibility, only splits with weights ≥0.01 are depicted. Polyploid taxa are indicated in bold. Series memberships within Sect. *Melampodium* are indicated as: CUP *Cupulata*, LEU *Leucantha*, LON *Longipila*, MEL *Melampodium*, SER *Sericea*. Scale bar represents mean edge weight. (From Weiss-Schneeweiss et al., 2009: 1179)

The inferences of ancestral base chromosome number of *Melampodium*, and pathways of chromosomal change within the genus, were done by McCann et al. (2016). The results confirmed that both ascending and descending dysploidy have played an important role in base chromosome number evolution. This study showed that inferring ancestral chromosome numbers within a group can be influenced by the reconstruction method used (maximum parsimony, maximum likelihood, Bayesian methods), branch length model (phylograms versus chronograms), and phylogenetic uncertainty (topological and branch length). Therefore, reliance on single trees, such as the Maximum Clade Credibility (MCC) tree, was discouraged, and instead, model averaging, with both phylogenetic and model uncertainty being taken into account, was recommended. The ancestral state reconstructions in *Melampodium* were performed using only the diploids (McCann et al., 2016;

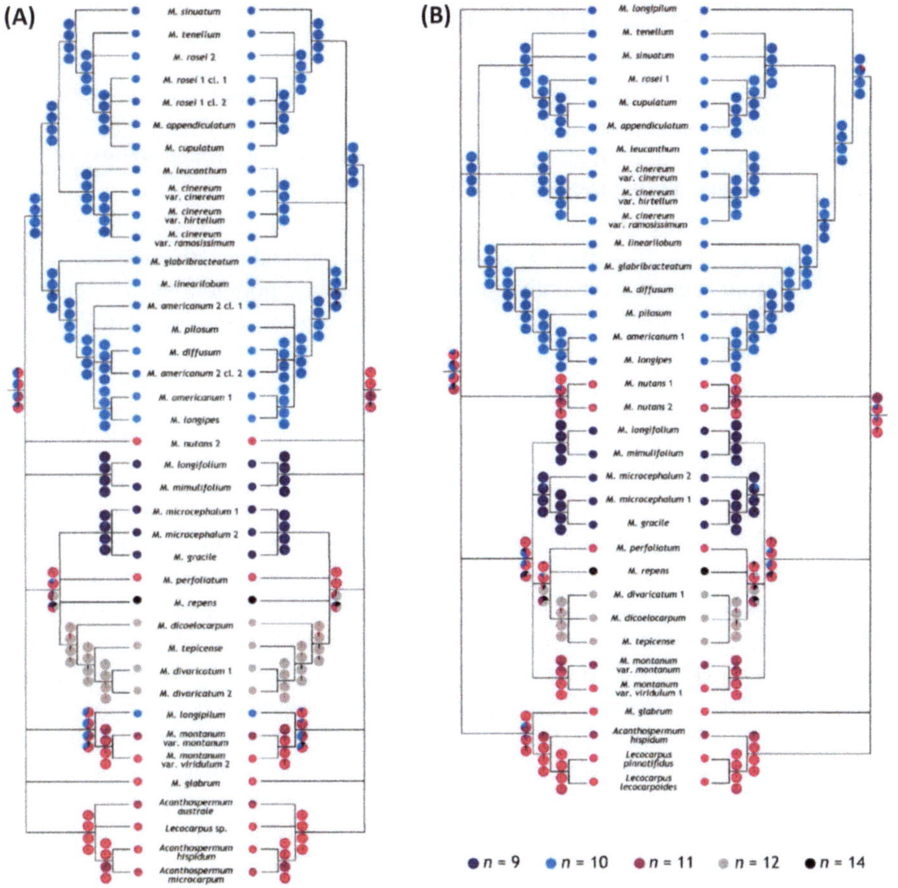

Fig. 7.3 Chromosome number reconstructions in *Melampodium* plotted on 95% majority rule-consensus trees from phylogenetic analysis of (**A**) nuclear sequence (ITS) data using BEAST (left) and MRBAYES (right) and of (**B**) plastid sequence (*mat*K) data using BEAST (left) and MRBAYES (right). At each node, the averages and, in case of maximum likelihood reconstructions, model-weighted probabilities of ancestral chromosome base numbers are shown (from top to bottom): ordered maximum parsimony (MP), maximum likelihood using CHROMEVOL (ML-CE), maximum likelihood using BAYESTRAITS (ML-BT), Bayesian Inference using Reversible Jump (BI-RJ). The pie charts represent the fraction of probability that is associated with a particular chromosome number. (From McCann et al., 2016: 8)

Fig. 7.3). The number $x = 11$ was inferred as ancestral in all analyses. From $x = 11$, the $x = 10$ lineage evolved by descending dysploidy. Origin of the $x = 12$ lineage was inferred to have occurred by ascending dysploidy, followed by another round of ascending dysploidy leading to the evolution of $x = 14$ ($x = 13$ has not been encountered in any species of the genus). Origin of the $x = 9$ lineages has been inferred to have involved descending dysploidy from ancestral $x = 11$.

Evolutionary Origin of Polyploids

Of particular significance in understanding chromosomal evolution within *Melampodium* is to infer the origin of the many exclusively polyploid species, 13 in all: *M. argophyllum* (*n* = 30), *M. aureum* (*n* = 33), *M. bibracteatum* (*n* = 28), *M. costaricense* (*n* = 24), *M. longicorne* (*n* = 30), *M. mayfieldii* (*n* = 20), *M. nayaritense* (*n* = 20), *M. northingtonii* (*n* = 24), *M. paniculatum* (*n* = 18, 27), *M. pringlei* (*n* = 30), *M. sericeum* (*n* = 30), *M. sinaloense* (*n* = 24), and *M. strigosum* (*n* = 20). In addition, four species have both diploid and tetraploid populations: *M. cinereum* (*n* = 10, 20), *M. dicoelocarpum* (*n* = 12, 23), *M. leucanthum* (*n* = 10, 20), and *M. repens* (*n* = 14, 28).

Autopolyploids

The origin of species of angiosperms by autopolyploidy has not been documented as frequently as via allopolyploidy, but several instances are known (Parisod et al., 2010). The major obstacle the autopolyploids face is to establish regular bivalent pairing in meiosis, hence restoring fertility. Within *Melampodium*, three species potentially originated via autopolyploidy: *M. aureum* (*n* = 33), *M. bibracteatum* (*n* = 28), and possibly also *M. argophyllum* (*n* = 30). *M. aureum* is an autohexaploid (*n* = 33) with the only close relative *M. montanum* with *n* = 11 (in Sect. *Rhizomaria*). A simple hypothesis, based on existing evidence, would be that *M. aureum* was derived by autopolyploidy from diploid *M. montanum*, perhaps also involving not-yet discovered (or no longer extant) autotetraploids of this species. *M. bibracteatum* is a tetraploid at *n* = 28, and the only other species known in Sect. *Bibractiaria* is *M. repens*, which has been counted as *n* = 14 and 28 (Weiss-Schneeweiss et al., 2009). Because this taxon tends toward autopolyploid race formation, a similar and perhaps much earlier autopolyploid event might have occurred with another population that later diverged and speciated into *M. bibracteatum*. The origin of *M. argophyllum* is more complex. This hexaploid species is very similar to its close relatives, *M. leucanthum* and *M. cinereum* (in Ser. *Leucantha*), and they are both diploid with tetraploid chromosomal races. An allopolyploid origin via a natural cross between a diploid individual of one species and a tetraploid individual of another, accompanied by genome duplication, could be envisioned (Stuessy et al., 2004). Detailed studies on the white-rayed complex (Rebernig et al., 2012), however, point more to an autopolyploid origin for *M. argophyllum*, although allopolyploidy cannot be ruled out, especially because some intermixing with the diploid species could be detected (Rebernig et al., 2012).

The autopolyploid systems of *Melampodium* studied in most detail have been the tetraploids in *M. cinereum* and *M. leucanthum* in the white-rayed complex, Ser. *Leucantha* (Rebernig et al., 2012). The production of 4*x* individuals occurred in abundance in the easternmost part of the ranges of both species (Figs. 5.3, 5.4, and

5.5). Initial studies of the tetraploids revealed multivalent formation in meiosis (Stuessy, 1971b), suggestive of autopolyploidy. Also, no qualitative morphological differences were found between the races in either species, although in *M. cinereum* a quantitative difference has been noted (cf. Fig. 5.2). Genome sizes were measured for multiple individuals in several populations of all species of the white-rayed complex (Rebernig et al., 2012), and the total genome content of the tetraploids was found to be doubled over that of the diploids, as might be expected from recent autopolyploid origins (e.g., Ozkan et al., 2006; Eilam et al., 2009). The karyotype structure of the tetraploids closely resembled that of the diploids (Fig. 7.4), and rDNA loci numbers were doubled.

Allopolyploids

In contrast to autopolyploidy, numerous examples of speciation via allopolyploidy are known in angiosperms (Grant, 1981; Masterson, 1994; Otto & Whitton, 2000; Linder & Rieseberg, 2004; Rieseberg & Willis, 2007; Ainouche et al., 2009; Wood et al., 2009). Some likely allopolyploids exist in several taxonomic sections of *Melampodium*. Studies by Stuessy and Brunken (1979) have shown that the tetraploid *M. paniculatum* ($n = 18$) can be crossed with diploid *M. gracile* and *M. microcephalum* and form partially fertile F_1 hybrids. The results do not show conclusively that these two species were directly involved with the allopolyploidization; perhaps there was also another diploid species involved, such as *M. longifolium* or *M. mimulifolium* (or their ancestor), both diploids ($n = 9$) in Ser. *Zarabellia* of the same Sect. *Zarabellia*. It is clear, however, that *M. paniculatum* is not of simple autopolyploid origin. The other cluster of related diploid and tetraploid species involved *M. tepicense* ($n = 12$), and the related *M. sinaloense* ($n = 24$) and *M. northingtonii* ($n = 24$). The two latter tetraploids have been described as new species (Stuessy, 1970c; Turner, 1988), but they are relatively scarce and have not yet been investigated intensively. Preliminary data, however, suggest that the tetraploids have evolved by allopolyploidy involving the widespread diploid *M. divaricatum* and *M. tepicense* (yielding *M. costaricense*), between *M. divaricatum* and *M. tepicense* (in parallel yielding *M. sinaloense*), and between *M. tepicense* and *M. dicoelocarpum* (yielding *M. northingtonii*). This fascinating evolutionary complex needs further attention.

The most conspicuous and well-investigated examples of speciation by allopolyploidy in *Melampodium* occur in Sect. *Melampodium* ($x = 10$). Recent molecular cytogenetic and phylogenetic analyses (Fig. 7.5) have allowed greater precision in determining the parentage of these allopolyploids. Most allopolyploids occur in Ser. *Sericea*, where all species are either tetraploid or hexaploid. Outside of this series, only two other uniformly polyploid species in the section are known: *M. argophyllum* ($n = 30$), already discussed above, and *M. mayfieldii* ($n = 20$). This latter species is rare, found only near El Tuito in western Jalisco, and it is placed in Ser. *Melampodium*.

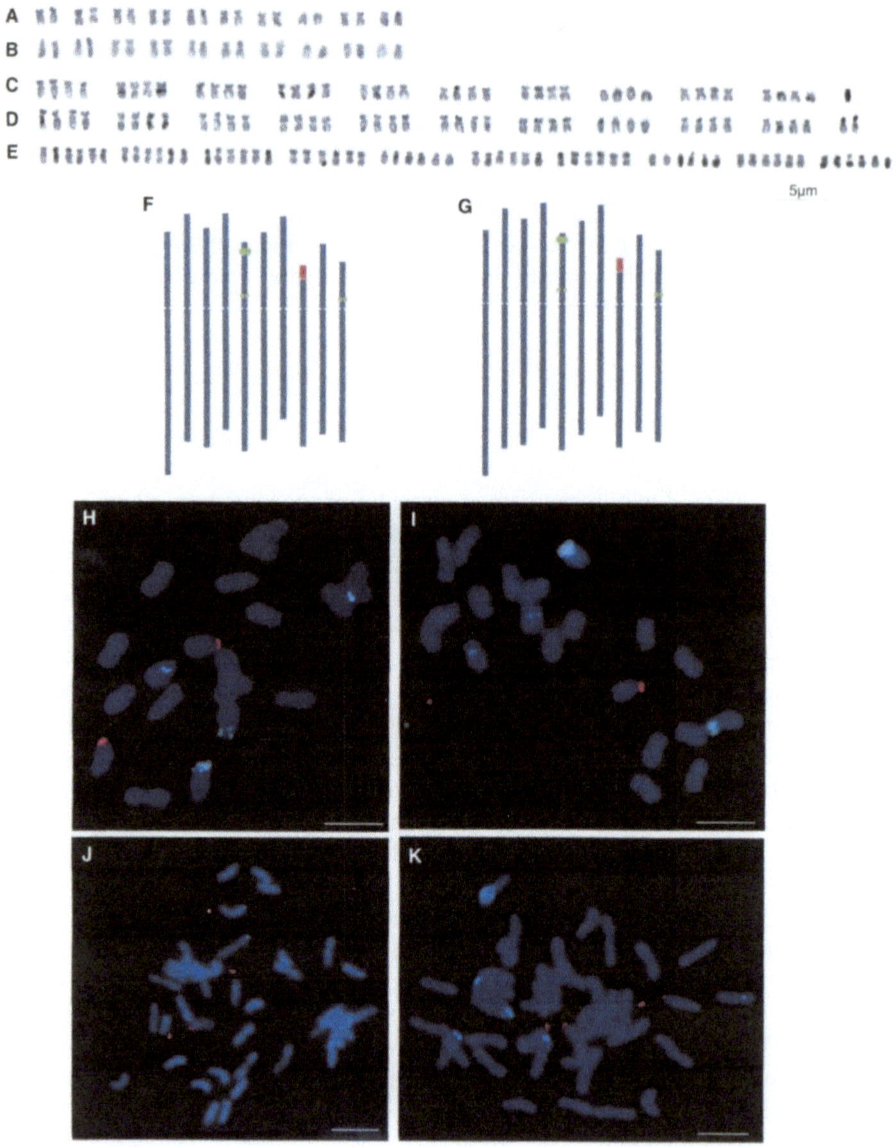

Fig. 7.4 Chromosomal analyses of the white-rayed species of *Melampodium*. Karyotypes (**A–E**) and idiograms (**F**, **G**) of *M. cinereum* var. *cinereum* (**A**, **F**, 2*x*, 18688; **C**, 4*x*, 18703), *M. leucanthum* (**B**, **G**, 2*x*, 19056; **D**, 4*x*, 18683) and *M. argophyllum* (**E**, 6*x*; 19059). **H–K** *In situ* chromosomal localization of 5S rDNA (green signals) and 35S rDNA loci (red signals) in diploid (18690; **H**) and tetraploid (18700; **J**) *M. cinereum* var. *cinereum*, as well as diploid (19056; **I**) and tetraploid (18683; **K**) *M. leucanthum*. Scale bars (**H–K**) = 5 μm. (From Rebernig et al., 2012: 1050)

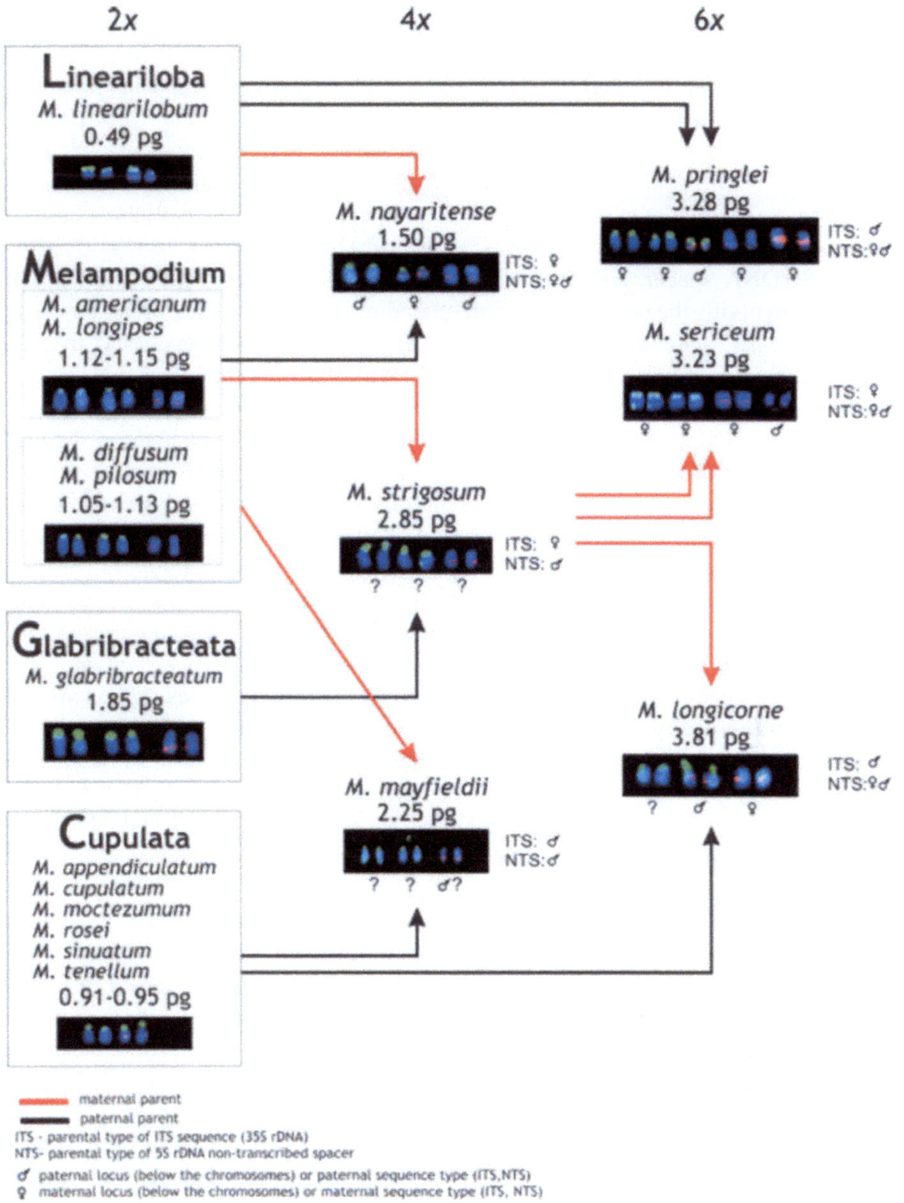

Fig. 7.5 Origin of and genome evolution in allopolyploids of *Melampodium* sect. *Melampodium*. Putative maternal and paternal lineages are indicated by red and black arrows, respectively. Genome size (1C), chromosomes carrying 35S (green signals), and/or 5S rDNA (red signals), and their putative provenance, as well as type of ITS and 5S rDNA non-transcribed spacer (NTS) sequences are shown. (From Weiss-Schneeweiss et al., 2012: 221)

Three species of Ser. *Sericea* have bimodal karyotypes (see Fig. 5.8C–E), which strongly suggests allopolyploid origins. These are *M. nayaritense* (*n* = 20; 4*x*), *M. pringlei* (*n* = 30; 6*x*), and *M. sericeum* (*n* = 30; 6*x*). All three species possess a set of ten small chromosomes, similar to those of *M. linearilobum* (*n* = 10; Fig. 5.7C), in addition to either one (in tetraploid) or two (in hexaploids) sets of larger chromosomes. Molecular phylogenetic investigations (Blöch et al., 2009) have suggested *M. linearilobum* to be involved with allopolyploidization in all three cases.

Subsequent phylogenetic analyses of several DNA markers (*mat*K and *psb*A-*trn*H, 5S rDNA spacer, and the two low copy nuclear genes, *Pgi*CI and *Pgi*CII; Fig. 7.6), involving the polyploids and all other diploid species of Ser. *Melampodium*, Ser. *Cupulata*, and Ser. *Sericea*, plus *M. glabribracteatum* alone in Ser. *Glabribracteata*, provided insights into the origin of all polyploids of this section. Three species were inferred to be of allotetraploid origin (*n* = 20): *M. nayaritense*, *M. strigosum*, and *M. mayfieldii*. *M. nayaritense* contained the genome of the diploids *M. linearilobum* on the one hand and *M. americanum* or *M. longipes* on the other hand, the latter two in Ser. *Melampodium* (Weiss-Schneeweiss et al., 2012; Fig. 7.6C–E). *M. strigosum* was also inferred to be derived from diploid progenitors in Ser. *Melampodium*, probably involving *M. americanum* or *M. longipes* (or their progenitors) and *M. glabribracteatum*. Additional evidence of this parentage comes from genomic *in situ* hybridization (GISH; Fig. 7.7), which reveals the presence of both parental diploid genomes of *M. glabribracteatum* and *M. americanum* in the karyotype of *M. strigosum* (Fig. 7.7A). The third tetraploid species, *M. mayfieldii*, resides taxonomically in Ser. *Melampodium*, having involved *M. diffusum* or *M. pilosum* as one diploid parent, but the other parent appears to have been from Ser. *Cupulata*; the latter connection is seen clearly in the phylogeny based on the analyses of low-copy genes (Fig. 7.6D, E) as well as with ITS and 5S rDNA (Fig. 7.6B, C). Which species might have been involved, however, could not be determined.

The origins of three hexaploid species (*n* = 30), all in Ser. *Sericea*, have been investigated: *M. sericeum*, *M. pringlei*, and *M. longicorne*. The first two species are very similar morphologically, such that Stuessy (1972) earlier suggested that with more study and collecting, the latter might be reduced to synonymy into the former. With recent collecting and molecular phylogenetic studies, it has become clear that the two are distinct species. Early cytological studies (Stuessy, 1970a) with *M. sericeum* had revealed two sizes of chromosomes in the hexaploid karyotype, 10 small and 20 large, which suggested an allopolyploid origin. Karyotypic (Weiss-Schneeweiss et al., 2009) and molecular phylogenetic studies (Blöch et al., 2009) have revealed *M. pringlei* to also have the same bimodal karyotype (Fig. 7.7C) and suggested that in both cases the small genome was contributed by *M. linearilobum*. Particularly diagnostic of this parentage have been results from analysis of low-copy gene *Pgi*C (Fig. 7.6D, E), further confirmed by mapping of parental DNAs onto the chromosomes of the polyploids using GISH, showing clearly the genome of *M. linearilobum* within that of *M. sericeum* (Fig. 7.7B) and *M. pringlei* (Fig. 7.7C). The other parental genome within both hexaploids has now been revealed as the tetraploid *M. strigosum* (Fig. 7.6), again being confirmed with GISH (Fig. 7.7B, C).

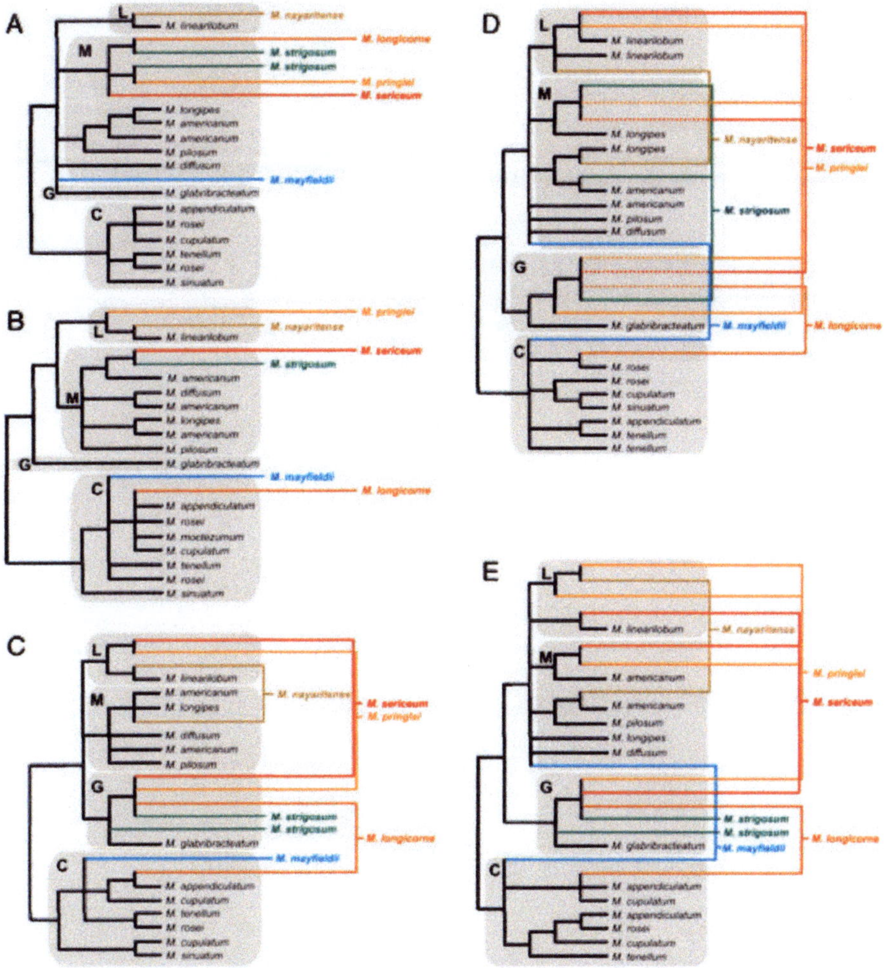

Fig. 7.6 Phylogenetic relationships of species of *Melampodium* Sect. *Melampodium* determined from sequences of (**A**) plastid *mat*K and *psb*A-*trn*H spacer, (**B**) nuclear ITS, (**C**) nuclear 5S rDNA NTS, (**D**) nuclear low-copy gene *Pgi*CI, and (**E**) nuclear low-copy gene *Pgi*CII. Different phylogenetic positions of homeologues from tetraploid (central lane names) and hexaploid taxa (right lane names) are the result of reticulate evolution involving different diploids (left lane names). Homeologues shared between tetraploids and hexaploids suggest involvement of the tetraploid in the origin of the hexaploid and are indicated by dashed lines. Gray rectangles indicate circumscription of phylogenetically and cytologically defined genomic groups (*C* Cupulata; *G* Glabribracteata, *L* Leariloba, *M* Melampodium). (From Weiss-Schneeweiss et al., 2012: 218)

These two origins of *M. sericeum* and *M. pringlei* from the same ancestors, but perhaps independently (they differ in number of 5S and 35S rDNA loci; Fig. 7.5), show the recurrent nature of allopolyploidy in angiosperm lineages. These allopolyploids formed relatively recently, perhaps within the past 1 million years (Fig. 7.8A) as evidenced by a dated phylogenetic network involving these same (and related) species.

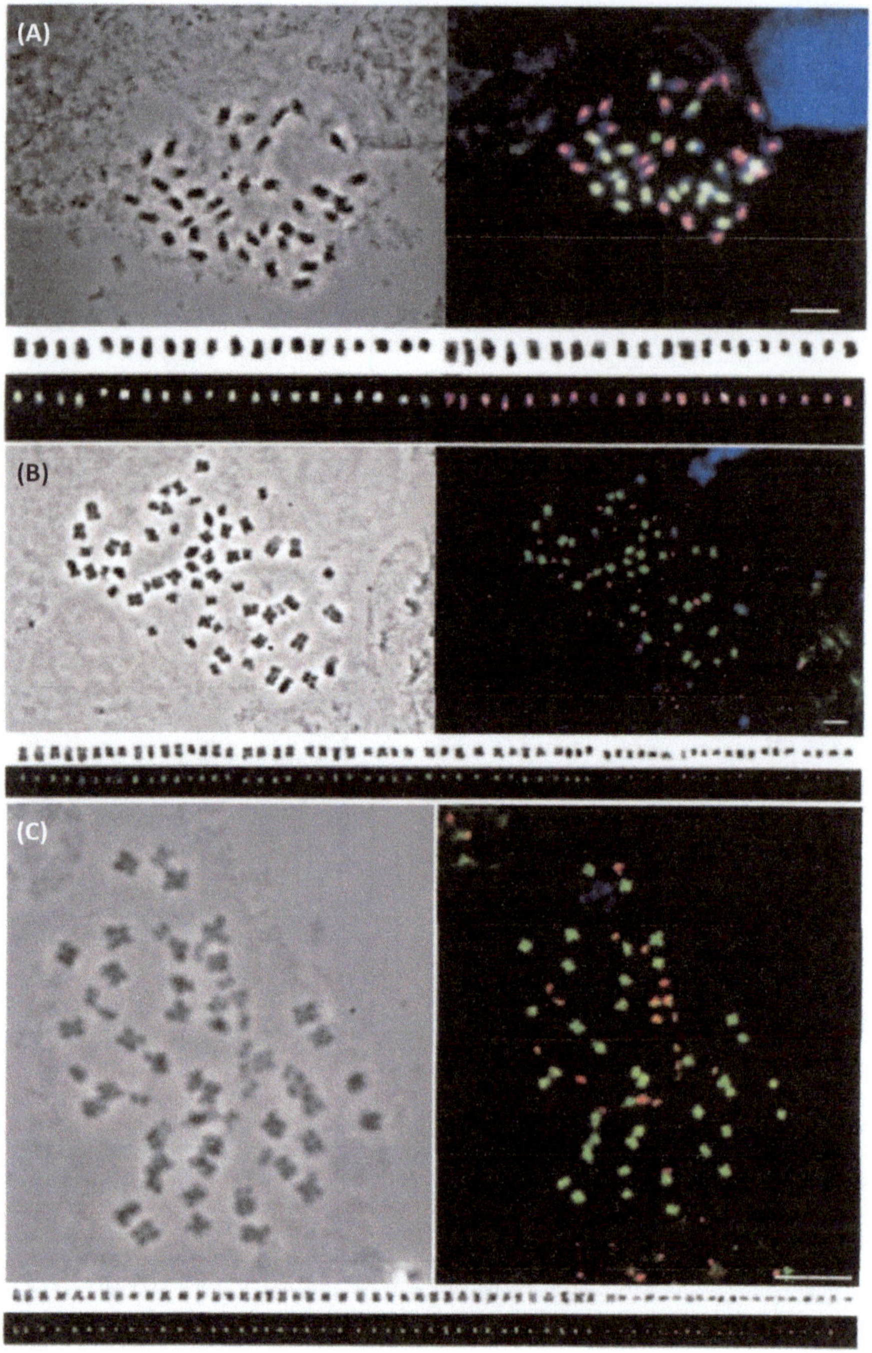

Fig. 7.7 Phase contrast chromosomes (unstained), GISH, and cut-out karyotypes on mitotic chromosomes of the three allopolyploids in *Melampodium*. (**A**) Allotetraploid *M. strigosum* with labeled genomic DNA of diploid *M. glabribracteatum* (green) and diploid *M. americanum* (red). (**B**, **C**) GISH of allohexaploid *M. sericeum* (**B**) and of allohexaploid *M. pringlei* (**C**) with labeled genomic DNA of diploid *M. linearilobum* (red) and allotetraploid *M. strigosum* (green). Scale bar = 5 µm. (From McCann et al., 2018: 1016)

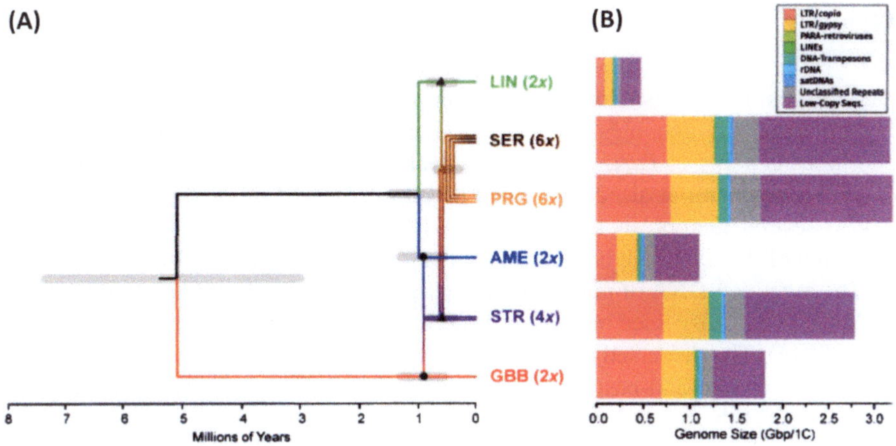

Fig. 7.8 Phylogenetic relationships and repeat composition in analyzed species of *Melampodium*. (**A**) Dated phylogenetic network showing relationships between allopolyploids and parental taxa for the single allopolyploid origin scenario. Divergence times are shown at the nodes with 95% highest-posterior density intervals indicated by the gray bars (cross-braced nodes are marked by dots and triangles). (**B**) Genomic proportions of repetitive elements for each species according to genome size and colored by repeat type (see legend). AME *M. americanum*, GBB *M. glabribracteatum*, LIN *M. linearilobum*, PRG *M. pringlei*, SER *M. sericeum*, STR *M. strigosum*. Ploidy level of each species is indicated as 2*x*, 4*x*, or 6*x*. (From McCann et al., 2018: 1015)

The hexaploid *M. longicorne* is also a member of Ser. *Sericea*, but its karyotype is not bimodal. More recent analyses of the karyotypes (Weiss-Schneeweiss et al., 2009) and molecular phylogenetic analyses (Fig. 7.6) revealed *M. longicorne* to have originated from the tetraploid *M. strigosum* in Ser. *Sericea* and one of the diploid species in Ser. *Cupulata*. This is of biogeographic interest, as *M. longicorne* occurs mainly in southeastern Arizona, which is adjacent to other species of Ser. *Cupulata*, particularly the diploid *M. appendiculatum* or even *M. moctezumum*.

Remarkable is the repeated involvement of the diploid *M. linearilobum* and the tetraploid *M. strigosum* in subsequent and recurrent polyploidization events. *M. strigosum* is involved with the origin of three hexaploid species, the similar *M. sericeum* and *M. pringlei*, and the morphologically different *M. longicorne*. Despite the obvious past hybridizations that have taken place in the origin of these tetraploid and hexaploid species, at present there is no evidence of active hybridization among them. In fact, there is very little hybridization known within the genus as a whole (see Chap. 8).

One of the central questions regarding the origin of allopolyploids in *Melampodium* is why it has taken place preferentially in this part of the genus. In general, a higher chance of allopolyploid formation seems to occur when there is high divergence between the parents (Chapman & Burke, 2007; Paun et al., 2009; but see Buggs et al., 2008, 2009, 2011; Stuessy & Weiss-Schneeweiss, 2019; Wagner et al., 2019), and also in arctic or weedy groups (Brochmann et al., 2004; Brysting et al., 2007; Kim et al., 2008), but beyond those possibilities, little is known about reasons for allopolyploidy developing strongly in some lineages and not in others (Soltis et al., 2010).

Genome Evolution

Nuclear genome size is known to vary considerably among flowering plants, ranging from the tiny genomes of *Genlisea* (Lentibulariaceae; ca. 0.06 pg/1C) to the enormous genome of *Paris japonica* (Melanthiaceae; 152.2 pg/1C) (Pellicer et al., 2010; Fleischmann et al., 2014). Such broad differences in genome sizes may be due to several factors, including polyploidy, but fundamental is also differential evolution of the repetitive component of the genome (Bennetzen, 2005).

The repetitive DNA of plant genomes exists in two broad categories (Bennetzen & Wang, 2014): dispersed mobile elements and tandem repeats (Weiss-Schneeweiss et al., 2015). The dispersed elements include mainly the DNA transposons and retrotransposons (cut-and-paste, and copy-and-paste elements, respectively). Repeats represented in the highest copy numbers in the majority of plant genomes are the long-terminal repeat (LTR) retrotransposons that are usually dispersed throughout the whole genome. These encompass two major superfamilies, Ty1-*copia* and Ty3-*gypsy* (Wicker et al., 2007), and they differ in predominance among plant genera and families (Novák et al., 2014; Kelley et al., 2015; Macas et al., 2015). Tandem repeats include ribosomal DNAs (5S and 35S) and satellite DNAs. In contrast to transposable elements, tandem repeats are usually located in one-to-many distinct loci in the genome; the only coding tandem repeats are ribosomal DNA loci. Satellite DNAs consist of long arrays of non-coding monomers located in (peri)centromeric, telomeric, and interstitial chromosomal regions. Satellite DNAs can make up a substantial portion of the plant genome, and they are also known to evolve quickly (Melters et al., 2013; Garrido-Ramos, 2015).

The rate and extent of removal and amplification of different repeats in the genome is one of the two major drivers of genome size evolution. The availability of high-throughput sequencing (next-generation-sequencing, NGS) and the RepeatExplorer pipeline (Novák et al., 2013, 2020) have now provided opportunities for comparisons to be made across species in different groups to begin to understand better this aspect of genome evolution. Furthermore, studies done thus far have revealed that repeat family abundances do contain phylogenetic information (Renny-Byfield et al., 2013; Macas et al., 2015; Dodsworth et al., 2016).

Repetitive DNA fraction (repeatome) evolution was investigated among the diploid species of *Melampodium* Sect. *Melampodium* (all $n = 10$; McCann et al., 2020). These genomes contain between 46–70% repeats, with larger genomes harboring more repetitive sequences than smaller ones. All species of the section were examined (13) except for two: *M. sinuatum*, a rare species growing only at the tip of Baja California (Villaseñor et al., 2011), and *M. longipilum* that was very phylogenetically divergent (see Fig. 6.6), and therefore it was deliberately excluded from these genomic comparisons. The objectives were to test the hypotheses that: (1) both up- and downsizing of genomes have occurred during evolution of the lineage; (2) repetitive DNA patterns would correlate with known phylogenetic patterns in the genus; and (3) repeat lineages have evolved at different rates.

A common post-polyploidization trend in genome evolution is genome downsizing, or less frequently upsizing (Weiss-Schneeweiss et al., 2006; Leitch & Leitch,

2008; Leitch et al., 2008). The genome sizes of allopolyploids in *Melampodium* Sect. *Melampodium*, however, were additive compared to the two putative parents (Fig. 7.6). It might be that these polyploid complexes in Sect. *Melampodium* Ser. *Sericea* are all of recent origin, certainly within only 1 million years (Fig. 7.8), and downsizing has just not yet taken place. Counterbalancing influences of loss of rDNA loci and genome expansion by retroelement activation (Bennetzen, 2005) may also have occurred. Phylogenetic analyses using BEAST and ancestral genome size reconstruction showed mixed patterns of increases and decreases in genome size within diploids of the group (McCann et al., 2016). Phylogenetic analysis using Ty1-*copia*, Ty3-*gypsy*, satellite DNAs, and all repeats taken together (Fig. 7.9), has

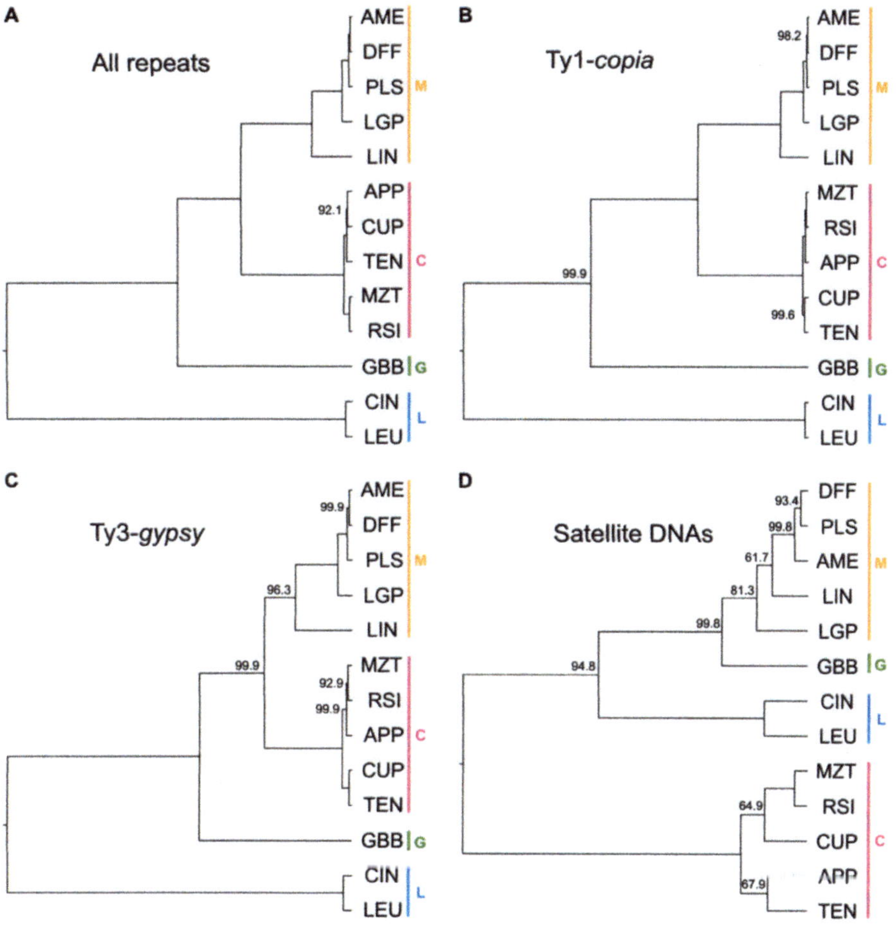

Fig. 7.9 Maximum a posteriori trees recovered from repeat-abundance-based phylogenetic analysis with a single rate parameter and a fixed root height of 1: (**A**) all repeats included with best model; (**B**) Ty1-*copia* elements only; (**C**) Ty3-*gypsy* elements only; and (**D**) satellite DNAs only. Color bars and capital letters next to species names indicate the taxonomic series to which species belong (C, Ser. *Cupulata*, pink; G, Ser. *Glabribracteata*, green; L, Ser. *Leucantha*, blue; M, Ser. *Melampodium*, yellow). (From McCann et al., 2020: 10)

shown patterns that resemble those seen with nuclear and plastid markers (ITS, *mat*K, etc.; see Chap. 6). The diploid species of Sect. *Melampodium* sort into four clearly defined taxonomic series: *Leucantha*, *Glabribracteata*, *Cupulata*, and *Melampodium*, and the use of several repeat types allows comparison regarding their rates of evolution. The SIRE lineage (*Ty1-copia*) was especially abundant in *M. glabribracteatum* and Ser. *Leucantha* (*M. leucanthum* and *M. cinereum*), suggesting that this is a recently active retroelement type.

Chapter 8
Modes of Speciation

That speciation in *Melampodium* has been accompanied by chromosomal change is one of the major themes of this book. The genus contains 13 different chromosome numbers from $n = 9$ to 33 on five different bases: $x = 9$, 10, 11, 12, and 14. Previous chapters have shown that from an ancestral base of $x = 11$, descending and ascending dysploidy, followed by polyploidy to the tetraploid and hexaploid levels, has been one of the major factors accompanying diversification within the genus. Nonetheless, chromosomal change is just one of the processes resulting in speciation, because environmental factors are also influential in keeping gene pools apart or encouraging their mixing. The challenge of the present chapter is to provide a broader overview of speciation within *Melampodium*, using all available data.

Isolating Mechanisms in *Melampodium*

Closely related to modes of speciation are isolating mechanisms. If no restriction of gene flow among individuals of an ancestral population occurs, no speciation can ensue. Some barriers to gene flow must develop so that genetic isolation, over time, can result in new species. Isolating mechanisms in plants are generally sorted into prepollination and postpollination factors (Table 8.1). Before pollination, several environmental and intrinsic factors can impede gene exchange and maintain isolation. One of the most common is spatial isolation, whereby two species simply cannot exchange genes because of the distance involved, even considering pollen dispersal by wind or insects. This type of isolation is normally reinforced by differences in habitat (ecological isolation). Also important are intrinsic aspects of the plants involved, such as mechanical (structural) differences in the flowers, different flowering times, and differences in reproductive systems (e.g., selfing vs. outcrossing). If pollination occurs between species, then a spectrum of genetic incompatibilities comes into play to maintain isolation. If fertilization occurs, then postzygotic

T. F. Stuessy et al., *Systematics, Ecology, and Chromosomal Evolution of* Melampodium *(Asteraceae)*, https://doi.org/10.1007/978-3-031-81995-7_8

Table 8.1 Classification of isolating barriers to gene exchange between angiosperm species

I. Prepollination (environmental and intrinsic) factors
A. Environmental factors
1. Spatial isolation
2. Ecological (habitat) isolation
B. Intrinsic factors
1. Morphological (floral) differences
2. Phenological differences (flowering time)
3. Pollinator selectivity
4. Highly selfing or asexual reproductive system
II. Postpollination (genetic) factors
A. Prezygotic: cross-incompatibility (pollen unable to germinate on stigma or pollen tube unable to grow to ovule; different chromosome numbers reflect this)
B. Postzygotic: reduced fitness in hybrid generations
1. Hybrid inviability or weakness
2. Hybrid floral isolation
3. Hybrid sterility
4. Hybrid breakdown in advanced or backcross generations

Based on Levin (2000: 11) and Crawford (2014: 112)

problems may appear in the hybrids. Further, if hybrids reproduce successfully and yield a new F_2 hybrid generation, these same genetic factors may then come into play again.

Several different isolating factors occur within *Melampodium*. This would hardly be a surprise due to the wide distribution of species of the genus, especially within environmentally diverse Mexico, and due to the wide array of chromosome numbers. Distributions of the species have been outlined in Chap. 4 (see also Appendix 4 for maps of all species), and it suffices to repeat that species in Mexico occur from the northwest portion of the country down to southern Chiapas. Many of these species simply never come into contact, and hence, hybridization would be unlikely; spatial isolation is completely effective in these instances. With some species, they can occur close to each other spatially (geographically), but there can be elevational differences that provide effective isolation (Fig. 8.1). It is obvious from the data in Fig. 8.1 that much variation exists in elevational preferences among species of *Melampodium*, and these differences have ecological correlations.

The ecosystems in Mexico are very diverse, and several climatic and ecological descriptive systems have been proposed to provide a guide to this environmental heterogeneity. We have relied on the systems of vegetation zones based on Leopold (1950) and Rzedowski (1978), which resulted in a useful categorization of seven zones, three temperate (pine-oak forest, mesquite grassland, desert), and four tropical (tropical evergreen forest, savanna, tropical deciduous forest, thorn forest). Figure 4.13 shows the occurrence of the species of *Melampodium* within the vegetation zones of Mexico and Central America, based on data from Stuessy (1972) and Turner (1988, 1993, 2007). The pattern reveals that the genus occurs in all seven

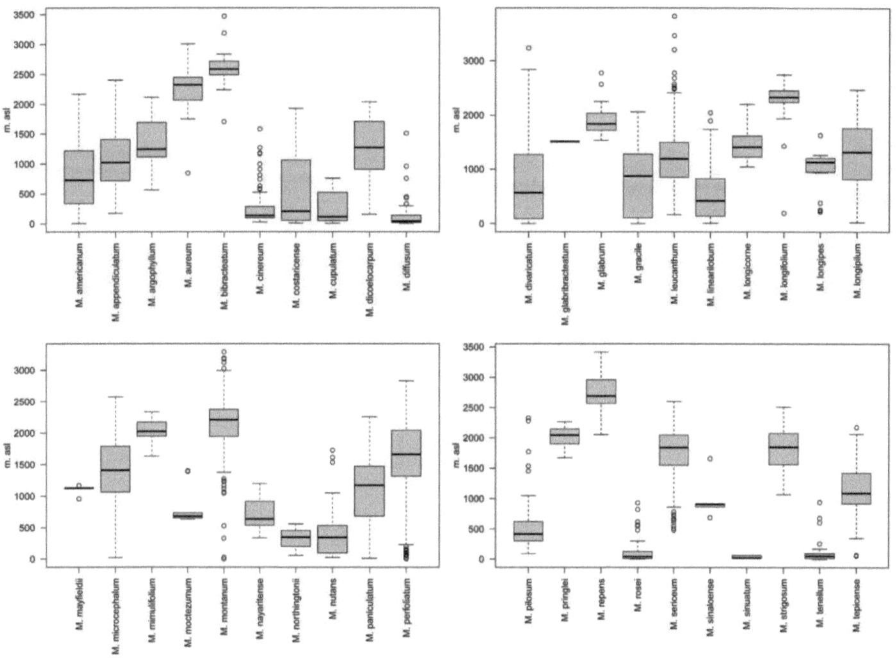

Fig. 8.1 Boxplots of variation in elevation for species of *Melampodium*. For statistics and methods, see Appendix 2

zones, with many species occurring in the pine-oak forest and tropical deciduous forest vegetation types, which also have elevational components. We presume that ecological and elevational zonation among species of the genus facilitate effective reproductive isolation. Other factors undoubtedly also play significant roles, such as soil type, humidity, bedrock geology, etc. Pollination ecology also can provide significant isolation in some groups of flowering plants (see examples in Faegri & van der Pijl, 1971), but not often in Asteraceae or among species of *Melampodium*. However, no rigorous study of pollinators in the genus has ever been done.

In some groups of plants, the time of flowering and type of breeding system can be powerful mechanisms for keeping species from hybridizing. Figure 8.2 shows the phenology of species in *Melampodium*. For collectors in Mexico during August and September, it comes as no surprise that numerous species of the genus are in flower at that time. There are some differences between taxa, but the generality is simply that phenology is not very effective in keeping species of *Melampodium* isolated.

The remaining isolating mechanisms are those that function after pollination has occurred, and these are genetic. A most important factor in *Melampodium* is prezygotic incompatibility, whereby pollination between species may take place via promiscuous insects but no fertilization occurs. This is where the different chromosome numbers among the species of *Melampodium* seem to provide a high level of genetic

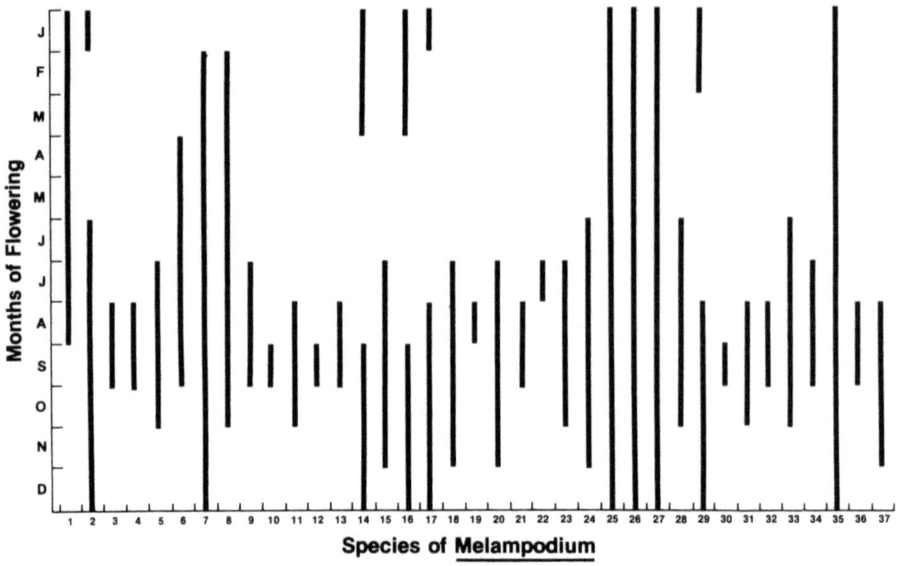

Fig. 8.2 Phenology of species of *Melampodium*. Numbers refer to species from the classification of Stuessy (1972; see Table 10.1). (From Sundberg and Stuessy, 1990: 84)

isolation. Some postzygotic isolation occurs, e.g., among artificial hybrids in Sect. *Zarabellia* (see below), but in general hybridization among species of the genus is not common. A descriptive model can be constructed (Sundberg & Stuessy, 1990) whereby species that have the same chromosome number usually have allopatric (non-overlapping) distributions. Figure 8.3 shows four species of Ser. *Cupulata* in northwestern Mexico, and they are all $n = 10$. They have distinct ranges, except for a small area of overlap between *M. cupulatum* and *M. rosei*, and in that area, morphological intermediates occur (Stuessy, 1972: 65), possibly suggesting hybridization. In contrast, Fig. 8.4 shows the distribution of five species of Ser. *Sericea*, with *M. strigosum* and *M. nayaritense* tetraploids ($n = 20$), and three hexaploid relatives at $n = 30$ (*M. sericeum*, *M. pringlei*, and *M. longicorne*). *M. strigosum* overlaps to a considerable extent the range of *M. sericeum* in central Mexico, but no hybrids are known. Likewise, *M. strigosum* overlaps completely the restricted distribution of *M. longicorne*, but without reported hybridization. The average degree of spatial overlap among species of these two examples is 14.11% among diploids and 45.6% among tetraploids and hexaploids (for methods for estimating area overlap, see Appendix 3).

During August and September in Mexico, it is not uncommon to find several species of *Melampodium* growing together in the same roadside population. A more complete survey of these co-occurrences is provided in Table 8.2, based on the field notebooks of Stuessy during the summers of 1965 and 1966. The most widely distributed species, *M. divaricatum* ($n = 12$), co-occurs with nine other species, but in

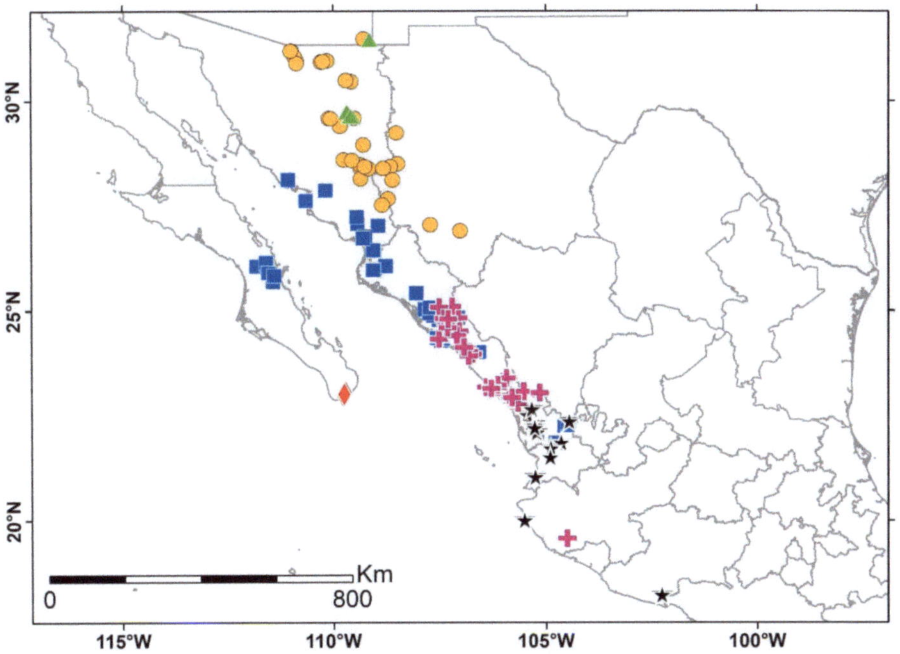

Fig. 8.3 Distributions of species of *Melampodium* in Sect. *Melampodium* Ser. *Cupulata*, all diploids, $n = 10$. *M. appendiculatum* (circles), *M. cupulatum* (squares), *M. moctezumum* (triangles), *M. rosei* (crosses), *M. sinuatum* (diamonds), *M. tenellum* (stars)

all cases, these have contrasting chromosome numbers ($n = 9, 10, 11, 18, 30$). Attempts to artificially cross *M. divaricatum* ($n = 12$; *Stuessy 547*) with *M. gracile* ($n = 9$; *Stuessy 537*) and *M. microcephalum* ($n = 9$; *Stuessy 681*) resulted in some seeds, but they were hollow inside (i.e., no embryo; Stuessy, unpubl.). Likewise, attempts to cross *M. gracile* ($n = 9$; *Stuessy 537*) with *M. rosei* ($n = 10$; *Stuessy 748*) resulted in no hybrid seeds (Stuessy, unpubl.).

Polyploid Speciation within *Melampodium*

In addition to speciation at the diploid level within closely related species complexes and the derivation by descending and ascending dysploidy of the different base numbers of $x = 9, 10, 11, 12$, and 14, polyploidy has been conspicuous as a speciation mechanism in several parts of the genus, and this has involved both autopolyploidy and allopolyploidy. Multiplying sets of chromosomes provides a high degree of reproductive isolation between the diploid progenitor(s) and the polyploid derivative. With autopolyploidy, the genome is simply completely multiplied, and in that sense, the polyploid is not genetically qualitatively different from the diploid, and hence is not typically accorded distinct specific status. Crossing between $2x$ and

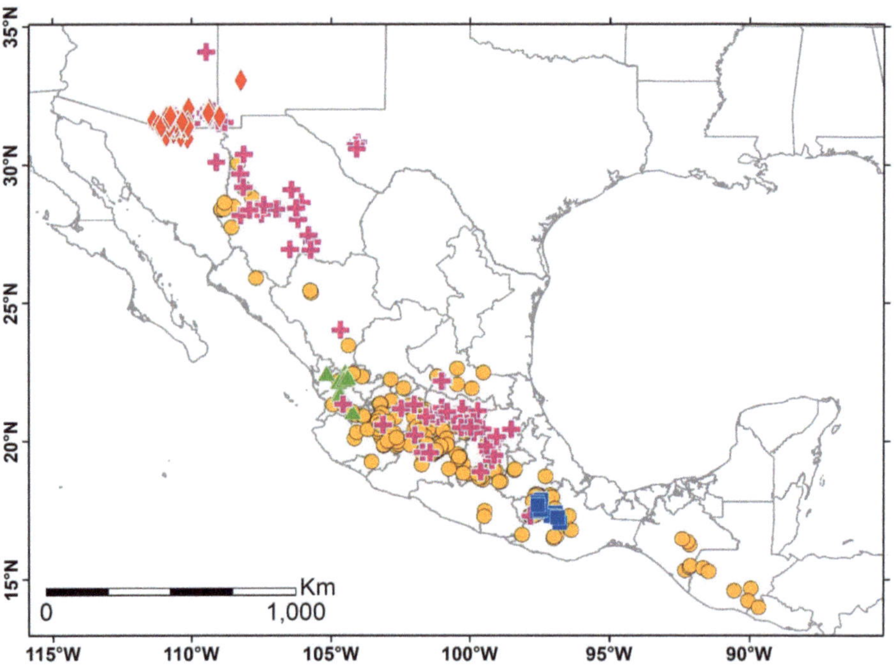

Fig. 8.4 Distributions of species of *Melampodium* in Sect. *Melampodium* Ser. *Sericea*, with tetraploids and hexaploids. *M. longicorne* (diamonds, $n = 30$), *M. nayaritense* (triangles, $n = 20$), *M. pringlei* (squares, $n = 30$), *M. sericeum* (circles, $n = 30$), *M. strigosum* (crosses, $n = 20$)

4*x* cytotypes has been documented in *M. leucanthum*, as evidenced by triploid plants (Rebernig et al., 2012; Figs. 5.3 and 5.5; Table 5.2; Appendix 1), but these individuals presumably are quite sterile.

Autopolyploidy

Autoploids have been discovered in several species of *Melampodium*. They have been documented extensively in *M. leucanthum* and *M. cinereum*, with both species possessing tetraploid cytological races in the eastern parts of their ranges. Diploid and tetraploid cytological races also occur in *M. repens*, $n = 14$ and 28, without any distinct morphological changes. Another very interesting example occurs in *M. dicoelocarpum*, which has $n = 12$ and $n = 23$ cytological races, morphologically indistinguishable. After initial autopolyploidization, the tetraploid has likely undergone descending dysploidy to the $n = 23$ level. Another case of infraspecific autoploidy occurs in the allotetraploid *M. paniculatum*, $n = 18$, with counts of $n = 27$ from populations in Brazil (Coleman, 1970), making it an autoallohexaploid.

One case exists in which a species may have evolved by autopolyploidization and become morphologically distinct: *M. aureum*, a hexaploid at $n = 33$. This is the

Table 8.2 Observed sympatry of species of *Melampodium* without known hybridization

	M. am.	*M. dic.*	*M. dic.*	*M. div.*	*M. gl.*	*M. gr.*	*M. lin.*	*M. lo.*	*M. lon.*	*M. mi.*	*M. pa.*	*M. pe.*	*M. ro.*	*M. se.*
	(10)	(12)	(23)	(12)	(11)	(9)	(10)	(10)	(10)	(9)	(18)	(11)	(10)	(30)
M. dicoelocarpum (12)	1													
M. dicoelocarpum (23)	–	–												
M. divaricatum (12)	2	–	1											
M. glabrum (11)	–	–	–	–										
M. gracile (9)	1	1	–	5	–									
M. linearilobum (10)	–	–	–	1	–	–								
M. longipes (10)	–	–	–	–	–	–	–							
M. longipilum (10)	–	–	–	1	–	–	–	–						
M. microcephalum (9)	1	–	–	1	–	–	–	–	–					
M. paniculatum (18)	–	–	–	3	–	–	–	–	–	–				
M. perfoliatum (11)	1	–	1	9	–	–	–	–	–	–	1			
M. rosei (10)	–	–	–	1	–	–	–	–	–	–	–	–		
M. sericeum (30)	2	–	1	9	2	1	1	1	–	1	–	5	–	
M. tenellum (10)	–	–	–	1	–	–	–	–	–	–	–	–	–	–

Data from field books of Stuessy from 1965 and 1966
From Sundberg and Stuessy (1990: 87)
Numbers in the data matrix refer to the number of times the pairs of species were found together; numbers in parentheses refer to haploid chromosome numbers
M. am. = *M. americanum*; *M. dic.* = *M. dicoelocarpum*; *M. div.* = *M. divaricatum*; *M. gl.* = *M. glabrum*; *M. gr.* = *M. gracile*; *M. lin.* = *M. linearilobum*; *M. lo.* = *M. longipes*; *M. lon.* = *M. longipilum*; *M. mi.* = *M. microcephalum*; *M. pa.* = *M. paniculatum*; *M. pe.* = *M. perfoliatum*; *M. ro.* = *M. rosei*; *M. se.* = *M. sericeum*

highest chromosome level known in the genus. This species is similar morphologically to *M. montanum*, but the former differs conspicuously in quantitative characters, especially in having larger flowering heads. *M. costaricense* was earlier hypothesized also to be of autopolyploid origin, but recent molecular data (Weiss-Schneeweiss et al., 2009) reveal it to be an allotetraploid (see Chap. 7).

Allopolyploidy

Allopolyploidization occurs normally via hybridization between two distinct species, accompanied by genome duplication. This results in a polyploid hybrid that combines two different genomes, and therefore, will have a unique combination of genetic and morphological traits, often sufficient for recognition as a distinct species. Furthermore, the allopolyploid will be reproductively isolated from the parents, because intermediate ploidy progeny tends to be sterile.

As has been explained previously, 11 species in *Melampodium* exist at the tetraploid or hexaploid level, excluding the putatively autohexaploid *M. aureum*, and these are likely of allopolyploid origin. *M. bibracteatum* (n = 28; 4x) is closely related to *M. repens*, which is known to have diploid and tetraploid races (n = 14, 28). The morphology of the former species, however, is quite different from the latter. The origin of the hexaploid (n = 30) *M. argophyllum* is still uncertain, but it obviously has evolved within the white-rayed complex. It may represent an original autohexaploid origin, as the tendency toward autopolyploid production exists in the related *M. leucanthum* and *M. cinereum*, but its morphology combines elements of the other two species and may be more indicative of allopolyploidy.

Species of definite allopolyploid origin have been documented in Sect. *Melampodium*, and particularly in Ser. *Sericea* (Fig. 8.5). Diploid species of Ser. *Melampodium*, Ser. *Glabribracteata*, and Ser. *Cupulata* have contributed genomes to yield the tetraploids *M. strigosum* (n = 20; 4x), *M. nayaritense* (n = 20; 4x), and *M. mayfieldii* (n = 20; 4x), and the hexaploids *M. sericeum* (n = 30; 6x), *M. pringlei* (n = 30; 6x), and *M. longicorne* (n = 30; 6x). These pathways of chromosomal evolution have been explained in detail in Chap. 7 and will not be repeated here. What is important for this chapter is to emphasize that allopolyploidy has been an important mode of speciation within the genus. This chromosomal variability, in consort with spatial and ecological isolation of populations, has been largely responsible for the high specific diversity within *Melampodium*.

Artificial Hybridizations in Sect. *Zarabellia*

In Sect. *Zarabellia*, due to close morphological relationships among the diploids (Stuessy, 1972), *M. gracile* and *M. microcephalum*, and the tetraploid (also counted as hexaploid in Brazil), *M. paniculatum* (Fig. 8.6A, C, E), it was decided to explore

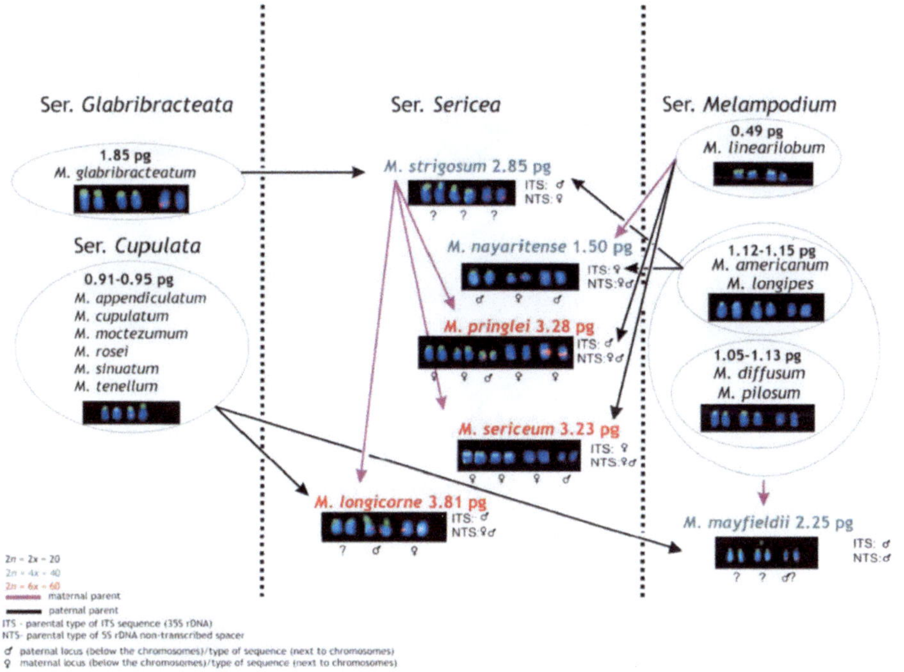

Fig. 8.5 Schematic representation of relationships among species and origin of polyploids in *Melampodium* Sect. *Melampodium*, showing ploidy levels, FISH signals, genome size, and type of ITS and 5S rDNA NTS sequences. Purple lines = paternal parents; black lines = maternal parents. (From Blöch, 2010: 120)

these relationships in more detail to understand the evolutionary origin of the tetraploid. A hypothesis had been advanced that at least one, if not both, of the diploids might be parents of the tetraploid (Stuessy, 1968), and hence an artificial crossing program was developed to test this idea. The objective was to infer genetic relationships among the species from data on crossability, pollen viability in F_1 hybrids (if they could be produced), and analysis of chromosomal pairing relationships during meiosis in the hybrids. These last data can offer clues to genomic compatibilities among the species. Furthermore, if interspecific diploid, homoploid hybrids might be successfully produced, it was hoped that through colchicine treatment, an artificial allopolyploid might be generated that could then be compared with the tetraploid species in all aspects.

The tetraploid, *M. paniculatum*, occurs primarily in Guatemala, El Salvador, Honduras, and Costa Rica (with some populations known in southernmost Mexico), whereas the diploid species are common in central and southern Mexico. It seemed plausible, therefore, that the two diploid taxa hybridized where they overlapped in southern Chiapas, and accompanied by genome duplication, yielded the allotetraploid *M. paniculatum*, which migrated southward into Central America away from competition with the parents. The distributions of *M. gracile* and *M. microcephalum*

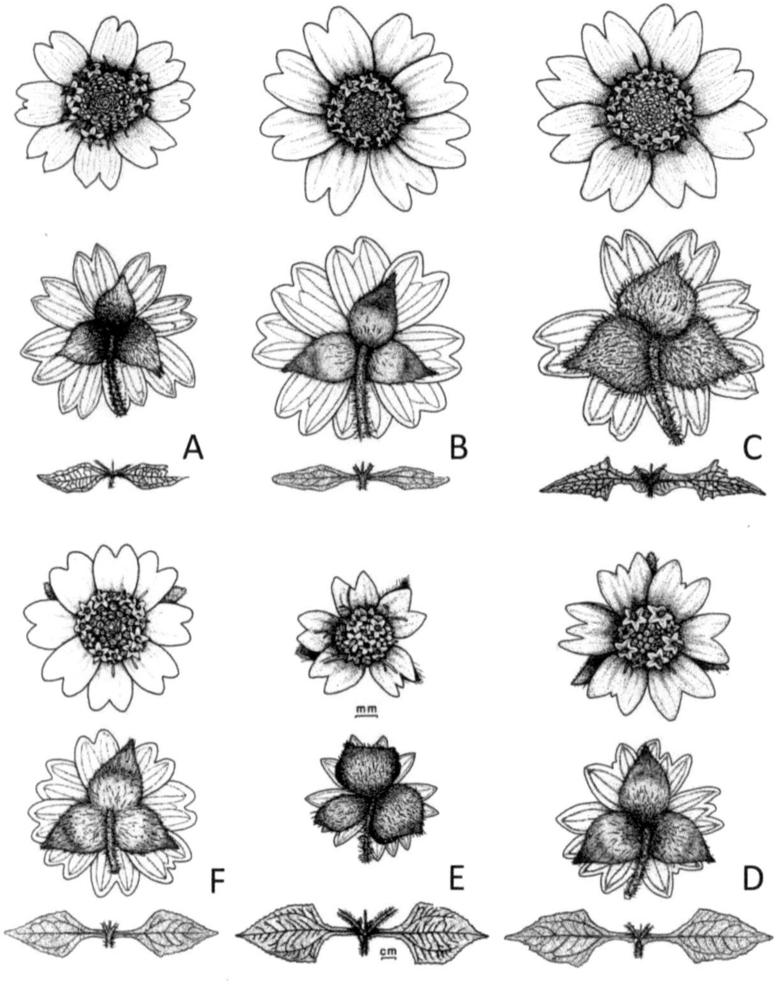

Fig. 8.6 Representative heads (top and bottom views) and mature leaves of species and F_1 hybrids in *Melampodium* Sect. *Zarabellia*. (**A**) *M. microcephalum* (*Stuessy 681*); (**B**) *M. gracile* × *M. microcephalum*; (**C**) *M. gracile* (*Stuessy 537*); (**D**) *M. gracile* × *M. paniculatum*; (**E**) *M. paniculatum* (*Stuessy 579*); (**F**) *M. paniculatum* × *M. microcephalum*. (From Stuessy and Brunken, 1979: 54). All same scale

in central Mexico show considerable overlap (Appendix 4), and it is here that occasionally intermediate morphologies are found, perhaps indicating more recent hybridization. These species are quite tolerant of disturbance (as with most species of the genus), often frequenting roadsides. Perturbation of the environment in central Mexico might have provided hybridization of the habitat (Anderson, 1948) and been a stimulus for interspecific hybridization.

Artificial hybridizations in *Melampodium* from a practical point of view are easy to accomplish. The set of sterile disc florets may be removed with fine forceps before anthesis, which renders the head functionally female. Tests were done to ascertain if such surgical maneuvers might interfere with the seed set, but this was not the case. Pollen placed on stigmas of ray florets resulted in fertile seeds. Tests were also carried out for autonomous apomixis in both parents and hybrids: emasculated heads were covered with lens-paper bags, and no seeds were produced (40 florets/3 plants in *M. gracile*; 38/3 in *M. microcephalum*; and 23/3 in *M. paniculatum*; Stuessy & Brunken, 1979). Tests for self-compatibilities were done on all three species, which revealed the highest levels in *M. paniculatum* (121 out of 151 florets, 80.1%), but less so in *M. gracile* (80/278, 28.8%) and *M. microcephalum* (30/200, 15.0%). These results indicated the importance of emasculation of the heads (i.e., removal of the disc) before anthesis, and covering them with small lens-paper bags (to exclude any incoming pollen from insects in the glasshouse) before interspecific crosses were attempted. During the crossing program, pollen from a donor head (which was allowed also to self-fertilize) was dusted on receptive female heads on three subsequent days to insure successful pollination.

The results from the crossing program are shown in Fig. 8.7. Seeds from one population of each species were selected for growing and use in the experimental crossings: *M. gracile*, Yucatán, 15 km N of Mérida, *Stuessy 537*; *M. microcephalum*, 5 km SE of Ciudad Hidalgo, Michoacán, *Stuessy 681*; *M. paniculatum*, Guatemala, 3 km S of Huehuetenango, *Stuessy 579*. Infraspecific crosses within each species were initially accomplished to provide a baseline of what levels of seed set might be expected within interspecific hybrids. Infraspecific levels of seed set were: *M. microcephalum*, 45.9%, *M. gracile*, 72.7%, and *M. paniculatum*, 73.2%.

Crosses among species were highly successful in both directions, with both male and female heads being used in each case (Fig. 8.7). Crossing occurred at a very high level between *M. microcephalum* and *M. paniculatum* with approximately 79% seed set in both directions, which is more than within each species. The crosses between the other diploid, *M. gracile*, and *M. paniculatum*, gave a slightly lower seed set (64.3% and 53.0%), but still high for an interspecific cross. In crosses between the diploids, *M. gracile* and *M. microcephalum*, when the former served as the male donor, only 23.5% seed set was observed, but with the latter as the pollen donor, 73.8% resulted. Checks were made on hybrid seeds to ascertain if they were full and viable (with tetrazolium chloride), and this was the case. These results confirmed that these three species are very closely related genetically, with the closest tie between *M. microcephalum* and *M. paniculatum*, which strongly suggested that this diploid, at least, was one of the parents in the allopolyploid origin of the tetraploid.

Hybrid seeds (236 total) were successfully germinated in the greenhouse and robust hybrid plants resulted. The percentages of hybrid seeds that germinated and survived to maturity (i.e., produced flowering heads) are shown in Table 8.3. The average cross (seed set, in both directions) between *M. microcephalum* and *M. paniculatum* was 29.5%, between *M. gracile* and *M. paniculatum* 14.5%, and between the diploids 8.5%. The data once again emphasize the close genetic tie

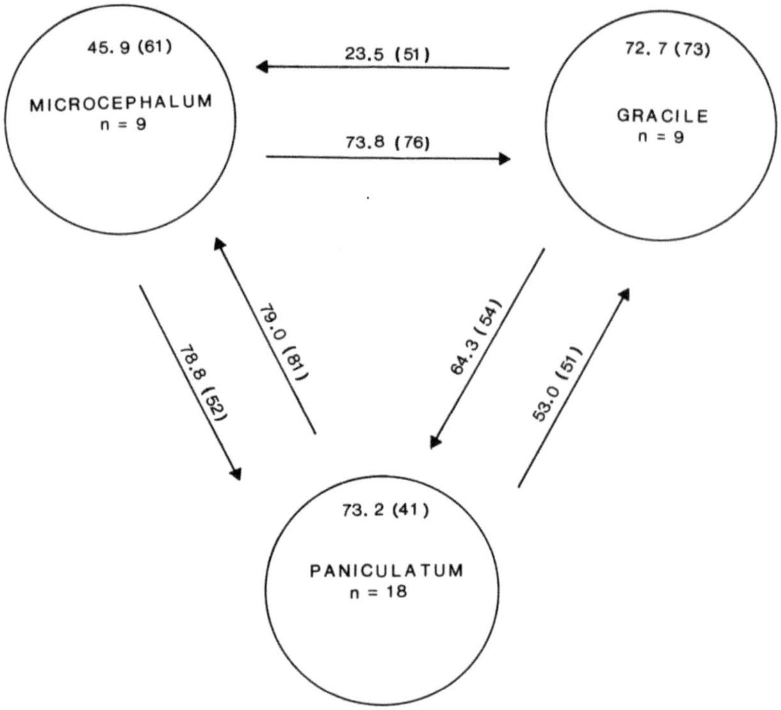

Fig. 8.7 Summary of crossing relationships within and among *Melampodium gracile* ($n = 9$), *M. microcephalum* ($n = 9$), and *M. paniculatum* ($n = 18$). Numbers in circles indicate percent seed obtained in infraspecific crosses. Arrows and numbers indicate percent seeds obtained in interspecific reciprocal crosses (arrows point toward female parent). Numbers in parentheses give the total number of florets involved with each type of cross. (From Stuessy & Brunken, 1979: 57)

Table 8.3 Survival and fertility in the F_1 generation from interspecific crosses among *M. gracile* ($n = 9$), *M. microcephalum* ($n = 9$), and *M. paniculatum* ($n = 18$)

	Survival data		Pollen stainability		Seed set (# hybrids/ % seed set)	
Hybrid (♀ x ♂)	# Seeds produced	% Survival to maturity	# Plants examined	% Stained (range)	Selfed	Crossed
microcephalum × *paniculatum*	64	36	36	12 (6–25)	7/0	14/0
paniculatum × *microcephalum*	41	23	13	10 (5–20)	5/0	10/0
paniculatum × *gracile*	36	19	5	6 (2–10)	7/0	7/0
gracile × *paniculatum*	27	10	9	6 (4–10)	4/0	4/0
gracile × *microcephalum*	56	17	13	11 (6–25)	7/0	5/3.6
microcephalum × *gracile*	12	0	–	–	–	–

From Stuessy and Brunken (1979: 58)

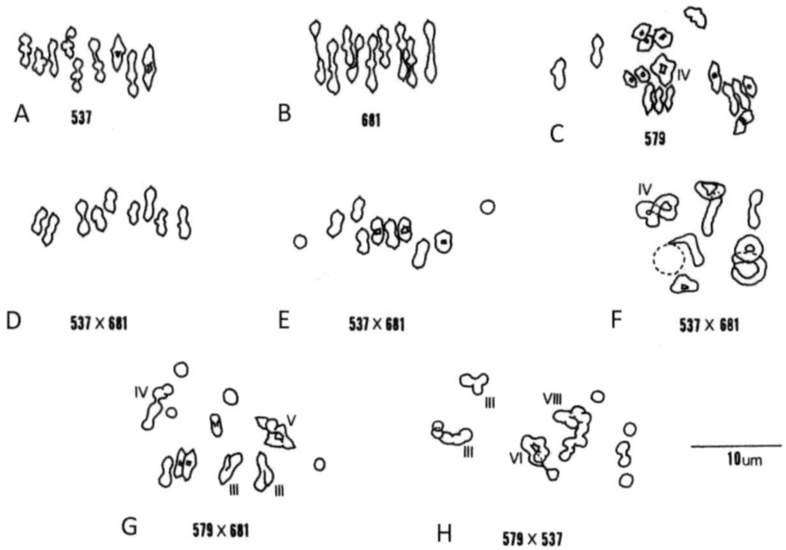

Fig. 8.8 Camera lucida drawings of meiotic chromosomes of species and hybrids of *Melampodium* Sect. *Zarabellia*. (**A**) *M. gracile* (*n* = 9; *Stuessy 537*); (**B**) *M. microcephalum* (*n* = 9; *Stuessy 681*); (**C**) *M. paniculatum* (*n* = 18 [2*n* = 16II +1IV]; *Stuessy 579*). (**D–F**) *M. gracile* × *M. microcephalum* (2*n* = 2*x* = 18); (**D**) Metaphase I, nine bivalents; (**E**) metaphase I, eight bivalents and two univalents; (**F**) diakinesis, seven bivalents and one quadrivalent; (**G**) *M. paniculatum* × *M. microcephalum* (2*n* = 3*x* = 27), metaphase I, one pentavalent, one quadrivalent, two trivalents, four bivalents, and four univalents; (**H**) *M. paniculatum* × *M. gracile* (2*n* = 3*x* = 27), metaphase I, one octavalent, one hexavalent, two trivalents, one bivalent, and five univalents. (From Stuessy & Brunken, 1979: 56). All same scale

between *M. microcephalum* and the tetraploid, in contrast to the weaker affinity with *M. gracile*.

Analysis of meiotic behavior in the artificially produced hybrids added additional insights for relationships among the three species. Meiosis in the diploids is normal with 9 bivalents (Fig. 8.8A, B), but in the tetraploid, although normally 18 bivalents were observed, occasionally one quadrivalent was seen (Fig. 8.8C). In the cross between the two diploids, meiosis was generally normal with nine bivalents (Fig. 8.8D), but in 12 of 50 meiocytes observed in metaphase I, two univalents (Fig. 8.8E) were also seen. One quadrivalent occurred in two of the 50 meiocytes observed in diakinesis. Chromosomal pairing in the triploid hybrids showed much more variation, with various combinations of univalents, bivalents, trivalents, and higher order multivalents (Fig. 8.8G, H). Disjunction and tetrad formation were very irregular. Pollen stainability in all hybrid plants was low, 2–25% (Table 8.3), with considerable variation within each hybrid cross category; parental species pollen stainability was 80–95%. The lowest level of pollen stainability was in hybrids between *M. gracile* and *M. paniculatum* (2–10%).

It is important to emphasize that although efforts were made to create an F_2 hybrid generation, only two seeds were obtained from 262 florets involved with the crosses. This suggests that in nature, despite the relative ease of obtaining interspecific hybrids between these three species, and especially between the two diploids, the probability of having advanced hybrid generations and/or introgressants would be very limited. That the diploids can easily cross and produce mature hybrids suggests that some of the intermediate herbarium specimens encountered during the revision of the genus (Stuessy, 1972) might be F_1 diploid hybrids. This would be especially the case in the zone of considerable contact between the two species in central Mexico, which might have been encouraged by the disturbance of habitats through human activities.

As for the origin of *M. paniculatum* from one or both diploid species, the cytogenetic and molecular data taken together indicate that *M. microcephalum* and *M. gracile* are likely parents. The other two species in the monophyletic Sect. *Zarabellia*, *M. longifolium* and *M. mimulifolium*, are also $n = 9$. The morphology of the former is not particularly close to the three species investigated cytogenetically, but *M. mimulifolium* is. At the time these crosses were completed, *M. mimulifolium* was only known from the type locality, and it had not been rediscovered. In 2005, however, one additional population also in Oaxaca was found (*Stuessy* et al. *18656*), determined to be diploid ($n = 9$), and with morphology that fits well into this complex.

Chapter 9
Biogeography

Understanding biodiversity requires not only investigating phylogenetic relationships among species within a genus, plus understanding evolutionary processes leading to their origins, but also interpreting temporal and spatial affinities. Due to the recent molecular phylogenetic analyses in *Melampodium* (Chap. 6), it now is possible to re-examine spatial and temporal origins of the group. The geographic range of the genus is broad, from arid regions of the southwestern United States to higher elevation tropical habitats of northwestern South America, and with two disjunct species in southeastern Brazil. The concentration of species is clearly within Mexico (see maps in Chap. 4 and Appendix 4).

Geography of Species Diversity and Its Causes

A useful beginning for understanding geographic aspects of species diversity in *Melampodium* is to examine the number of species of the genus that occur in all states of Mexico as well as Central America. Table 9.1 provides the list of species and their frequency of occurrence in the states and ecoregions of Mexico (Olson et al., 2001). The species most broadly distributed are *M. divaricatum* (in 22 states, and 19 ecoregions), *M. perfoliatum* (17, 12), *M. sericeum* (15, 9), *M. gracile* (12, 7), *M. americanum* (11, 13), and *M. microcephalum* (10, 11). The map in Fig. 9.1 and Table 9.2 show the number of species of *Melampodium* that occur in the states of Mexico. The states with maximum species diversity are Michoacán (21 species), Oaxaca (20), Jalisco (18), México (16), and Guerrero (14). In Central America (Fig. 9.2), Guatemala has 11 species, but then the number drops southward toward Panama (only 3). Much of the diversity in this region is due to the broader tolerances of *M. divaricatum* and *M. linearilobum*, plus the stronger appearance of *M. paniculatum* in the region.

T. F. Stuessy et al., *Systematics, Ecology, and Chromosomal Evolution of* Melampodium *(Asteraceae)*, https://doi.org/10.1007/978-3-031-81995-7_9

Table 9.1 Occurrence of species of *Melampodium* in the states and ecoregions of Mexico

Species	Number of states	Number of ecoregions
Section *Rhizomaria*		
M. montanum		
var. *montanum*	1	2
var. *viridulum*	9	8
M. aureum	4	3
Section *Glabrata*		
M. glabrum	3	2
Section *Zarabellia*		
Ser. *Zarabellia*		
M. longifolium	6	4
M. mimulifolium	1	2
Ser. *Tribracteata*		
M. microcephalum	10	11
M. gracile	12	7
M. paniculatum	1	4
Section *Alcina*		
M. perfoliatum	17	12
Section *Bibractiaria*		
M. bibracteatum	4	2
M. repens	4	3
Section *Serratura*		
M. divaricatum	22	19
M. costaricense	0	0
M. northingtonii	1	1
M. tepicense	4	3
M. sinaloense	1	2
M. dicoelocarpum	4	4
Section *Nutantia*		
M. nutans	4	3
Section *Melampodium*		
Ser. *Longipila*		
M. longipilum	5	6
Ser. *Cupulata*		
M. cupulatum	4	4
M. appendiculatum	3	3
M. moctezumum	1	1
M. sinuatum	1	1
M. rosei	2	3
M. tenellum	3	2
Ser. *Leucantha*		
M. leucanthum	3	4

(continued)

Table 9.1 (continued)

Species	Number of states	Number of ecoregions
M. cinereum		
var. *cinereum*	2	2
var. *hirtellum*	3	2
var. *ramosissimum*	1	1
M. argophyllum	2	1
Ser. *Glabribracteata*		
M. glabribracteatum	1	1
Ser. *Sericea*		
M. sericeum	15	9
M. pringlei	1	1
M. strigosum	7	6
M. nayaritense	2	1
M. longicorne	1	1
Ser. *Melampodium*		
M. americanum	11	13
M. diffusum	2	5
M. pilosum	2	3
M. longipes	3	3
M. linearilobum	7	8
M. mayfieldii	1	1

Ecoregions after Olson et al. (2001)
From geographic data in Stuessy (1972) and Villaseñor (unpubl. database). Revised classification follows that of Stuessy et al. (2011)

Understanding the reasons for high species diversity of *Melampodium* along the southern portions of Mexico must relate, at least partly, to environmental factors. Important features of Mexico are the numerous climatic, ecological, and vegetation zones (see Chap. 4). Table 9.3 gives the states of Mexico with the number of species of *Melampodium*, vegetation types, and maximum elevation. The state with the most vegetation zones (after Leopold, 1950; see map, Fig. 4.9) is Puebla (seven vegetation types), which also has the highest elevation (along with Veracruz because of Pico de Orizaba, the highest point in the country, 5636 m), along with Guerrero, Jalisco, Michoacán, San Luis Potosí, and Veracruz (average elevation 4089 m) with six vegetation zones each, giving high levels of species ($x = 13.7$). The next group (Chiapas, Hidalgo, Oaxaca, Sonora, Tamaulipas) each contains five vegetation types, an average of 10.4 species, with an average elevation of 3453 m. The states with four vegetation types have average species of 7.8 and elevation of 3232 m, those with three average 6.8 species and 3170 m, and those with only two average 3.7 species and 2407 m. There is a positive correlation, therefore, of species number with the number of vegetation types, average species diversity, and highest elevation; the higher the elevation, the greater the specific and vegetational diversity. There are exceptions, e.g., México, which reaches to 5413 m but harbors only three

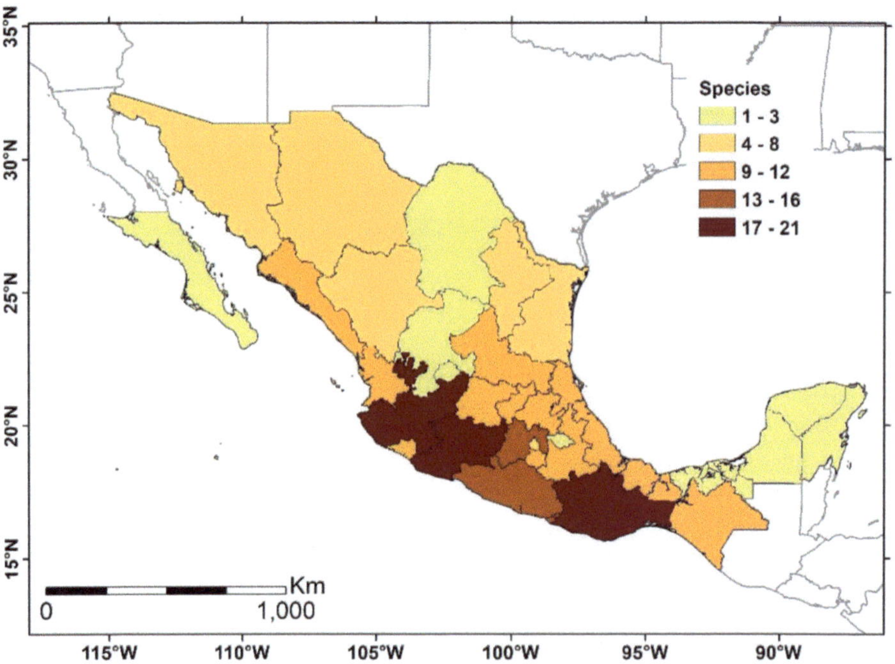

Fig. 9.1 Numbers of species of *Melampodium* occurring in the states of Mexico. Data from Villaseñor database (unpubl.). Michoacán and Oaxaca house the most species (21, 20)

vegetation zones. In general, elevation is a good indication of plant diversity within a region (e.g., Steinbauer et al., 2016).

Ancestral Habitat and Area

As a basis for reconstructions of both the ancestral habitat and the ancestral area of *Melampodium*, a consensus phylogeny was constructed manually from available phylogenetic information (Blöch et al., 2009; Stuessy et al., 2011; Weiss-Schneeweiss et al., 2012). We followed a conservative approach, collapsing areas of uncertain phylogenetic relationships to polytomies (due to poor resolution or contradictions). The phylogenetic position of *M. longipilum* differs irreconcilably among phylogenies inferred from plastid versus nuclear markers (Chap. 6); for analysis, we took the phylogenetic position inferred from plastid data as it agrees with morphological and karyological evidence. Species of allopolyploid origin (Chap. 7) were represented by each of their subgenomes (i.e., an allotetraploid appears twice on the tree, an allohexaploid appears three times), placing the two or three subgenomes together with the maternal or paternal diploid or, in the case of *M. pringlei* and *M. sericeum*, allotetraploid parental species. Assuming that the

Table 9.2 Ranking of states of Mexico by number of species of *Melampodium* contained within each. Distribution data from Villaseñor database (unpubl.)

State	Number of species
Michoacán	21
Oaxaca	20
Jalisco	18
México	16
Guerrero	14
Nayarit	12
Queretaro	12
Puebla	11
Chiapas	10
Colima	10
Guanajuato	10
Morelos	10
Sinaloa	10
Hidalgo	9
San Luis Potosí	9
Veracruz	9
Sonora	8
Chihuahua	6
Distrito Federal	6
Durango	6
Nuevo León	6
Tamaulipas	5
Coahuila	3
Yucatán	3
Zacatecas	3
Aguascalientes	2
Baja California Sur	2
Campeche	2
Tlaxcala	2
Quintana Roo	1
Tabasco	1

origin of the allopolypoid is at the site of the seed parent (i.e., the maternal parent), the subgenomes corresponding to the paternal parent of an allopolyploid were coded as "unknown" in subsequent analyses to avoid distortion due to duplication of species. Species of *Melampodium* were assigned from one to several of the following habitat types (following Sundberg & Stuessy, 1990): tropical evergreen forest; tropical deciduous forest; pine-oak forest; thorn forest; savanna, mesquite grassland; desert. Habitat evolution was reconstructed using maximum parsimony in Mesquite 3.70 (Maddison & Maddison, 2021). The ancestral habitat was inferred to have been pine-oak forests, the most frequently found habitat type in *Melampodium* (Fig. 9.3). Shifts to other habitat types mainly occurred along terminal branches (i.e., those leading to species), except for the ancestor of *M.* Sect. *Melampodium*

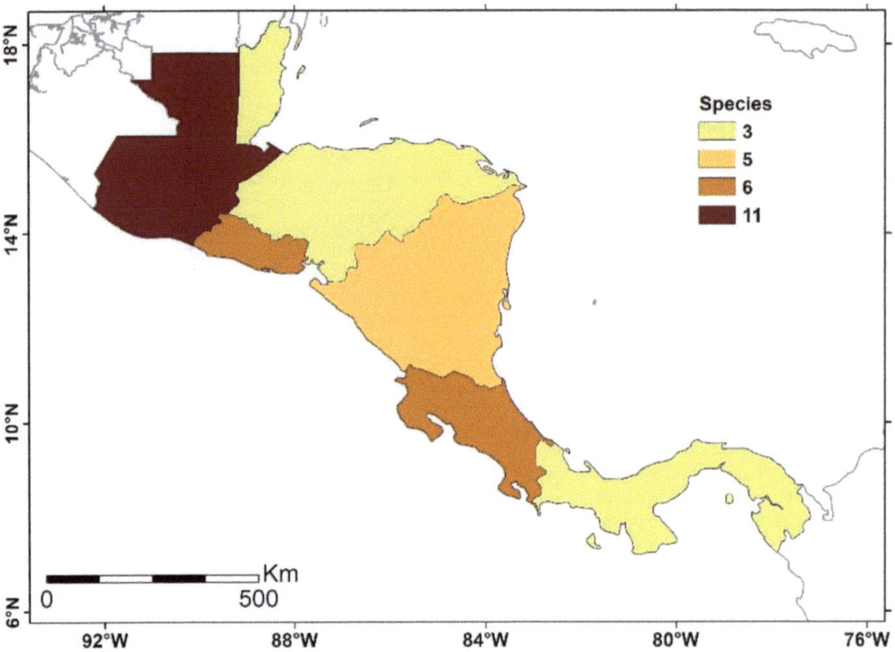

Fig. 9.2 Numbers of species of *Melampodium* occurring in the countries of Central America. (Data from Villaseñor database, unpubl.)

Ser. *Cupulata* and *M.* Sect. *Serratura*, whose ancestors were inferred to have inhabited tropical deciduous forests (Fig. 9.3). Despite numerous habitat shifts, transitions between habitat types are strongly restricted. As shown in Table 9.4, habitat changes essentially always were from pine-oak forests or tropical deciduous forests as source habitats, the only exception being mesquite grasslands, which was reached from a desert ancestor. These results suggest that ecological differentiation played no role in the early diversification of *Melampodium*, which took place almost exclusively in pine-oak forests.

Determination of the ancestral area of *Melampodium* followed techniques used with the ancestral habitat analysis described above. A data matrix of species within spatial regions of Mexico was prepared (Table 9.5) following the floristic units delimited by Villaseñor and Ortiz (2025). Reconstruction of ancestral areas with maximum parsimony in Mesquite 3.70 (Maddison & Maddison, 2021) is shown in Fig. 9.4. The results indicate that the West-Central region of Mexico was the ancestral area for *Melampodium*, which correlates with a high level of species diversity (Fig. 9.1). A shift to the Northwestern region of Mexico occurred in the ancestor of Sect. *Melampodium* (x = 10), the same region in which members of this section subsequently differentiated. As with the ancestral habitat analysis, the majority of geographic changes (e.g., extension into other parts of Mexico or into the United States) appear to have taken place in the context of species divergence.

Table 9.3 Association of number of vegetation types, number of species of *Melampodium*, and maximum elevation for states of Mexico

State	Number of vegetation	Number of species	Maximum elevation (m)
Puebla	7	11	5636
Veracruz	6	9	5636
Jalisco	6	18	4260
Michoacán	6	21	3840
Guerrero	6	14	3540
San Luis Potosí	6	9	3170
		$X = 13.7$	$X = 4089$
Chiapas	5	10	4067
Oaxaca	5	20	3700
Tamaulipas	5	5	3510
Hidalgo	5	9	3370
Sonora	5	8	2620
		$X = 10.4$	$X = 3453$
Colima	4	10	3820
Chihuahua	4	6	3303
Durango	4	6	3352
Zacatecas	4	3	3200
Sinaloa	4	10	2955
Nayarit	4	12	2760
		$X = 7.8$	$X = 3232$
México	3	16	5413
Morelos	3	10	5452
Nuevo León	3	6	3720
Coahuila	3	3	3715
Querétaro	3	12	3366
Baja California Sur	3	2	3095
Campeche	3	2	390
Yucatan	3	3	210
		$X = 6.8$	$X = 3170$
Tlaxacala	2	2	4438
Guanajuato	2	10	3366
Aguascalientes	2	2	3020
Distrito Federal	2	6	2240
Tabasco	2	1	1140
Quintana Roo	2	1	240
		$X = 3.7$	$X = 2407$

The West-Central region of Mexico includes the states of Colima, Guerrero, Jalisco, México, Michoacán, Morelos, and Tlaxcala. Significant is that three important mountain ranges converge here, providing extreme levels of habitat heterogeneity: the Trans-Mexican Volcanic Belt (TMVB), the Sierra Madre Occidental (SMOC), and the Sierra Madre del Sur (SMS). Between the first two zones is a

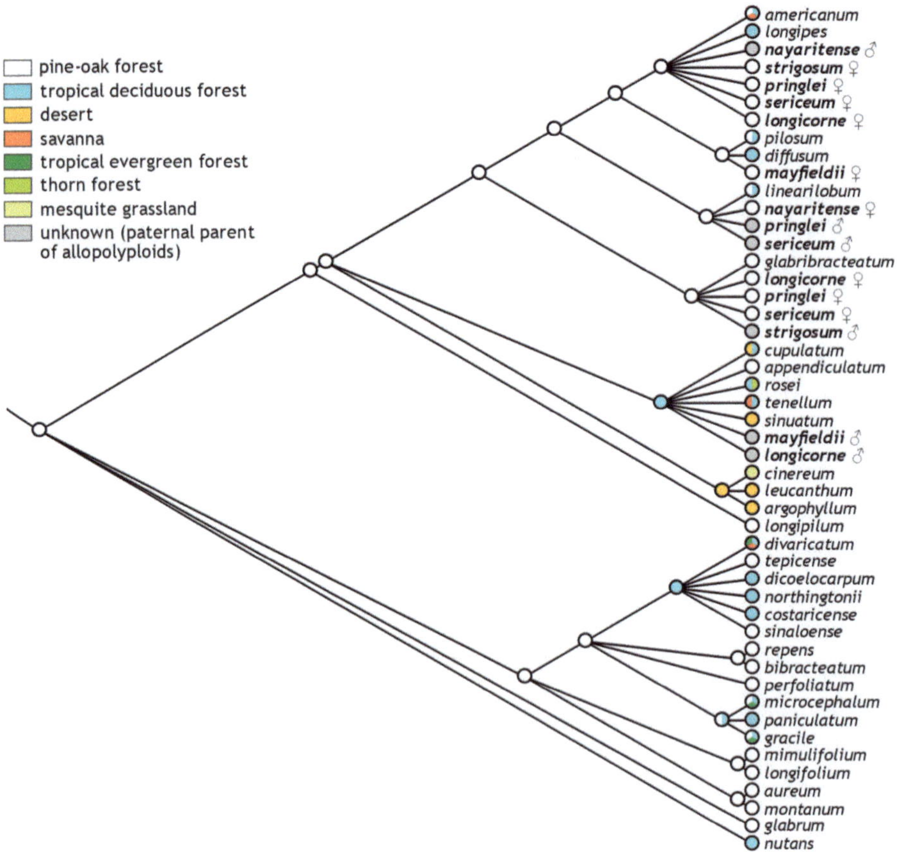

Fig. 9.3 Ancestral habitat reconstruction in *Melampodium*. Species in bold are allopolyploids represented by their subgenomes; male and female symbols refer to parentage

basin recognized as a floristic and biogeographic province, the Balsas River Basin (BAL). In addition, to the west of the Sierras is the Pacific Ocean Plain (PAC).

The TMVB is a volcanic chain located along the central portion of Mexico in a west-east orientation. Ferrusquía-Villafranca (2007) describes it as beginning in Nayarit and ending in Veracruz. This mountain range was formed in stages, first emerging in the west during the Miocene (15–17 million years ago), and with its eastern portion evolving more recently during the Pliocene and Pleistocene (about 7.4 million years ago; Moran-Zenteno, 1994; Becerra, 2005; Ferrari et al., 2012; Mastretta-Yanes et al., 2015).

The SMOC runs along the northwestern portion of Mexico, parallel to the PAC, from Sonora and Chihuahua to northern Jalisco, encompassing parts of the states of Aguascalientes, Durango, Nayarit, Sinaloa, and Zacatecas. To the north, it connects with the Rocky Mountains of the United States, and to the south with the TMVB. It resulted from volcanic episodes starting in the late Cretaceous and continuing until

Table 9.4 Average number of habitat shifts from source (in rows) to sink habitats (columns) among species of *Melampodium*

	Sink habitats						
Source	Pine-oak forest	Tropical deciduous	Desert	Savanna	Tropical evergreen	Thorn forest	Mesquite grassland
Pine-oak forest	–	7.59	1.00	0.35	0.35	0	0
Tropical deciduous	3.37	–	1.48	0.82	0.65	0.42	0
Desert	0	0	–	0	0	0	1.00
Savanna	0	0	0	–	0	0	0
Tropical evergreen	0	0	0	0	–	0	0
Thorn forest	0	0	0	0	0	–	0
Mesquite grassland	0	0	0	0	0	0	–

the Middle Miocene (Ferrari et al., 2007). In several of the states placed along this mountain range (e.g., Durango, Jalisco, Nayarit, and Sinaloa) a greater richness of species is found, consisting primarily of clades of Section *Melampodium*, several of them micro-endemics (e.g., *Melampodium longipes, M. mayfieldii, M. nayaritense, M. rosei, M. sinaloense, M. tenellum*, and *M. tepicense*).

The SMS is the third orographic province characterizing this West-Central region; it comprises the western part of the states of Guerrero and Oaxaca and geologically is characterized by rock formations dating back to the Mesozoic, as well as others resulting from volcanism that occurred during the Cenozoic (Santa María-Díaz, 2016). It is along these three mountain ranges where the temperate forests in the West-Central region of Mexico favored the origin and diversification of *Melampodium*. Among endemics in this mountain range can be mentioned *Melampodium glabribracteatum, M. mimulifolium*, and *M. northingtonii*.

The BAL has been largely shaped by orogenesis of the TMVB and the SMS. Both mountain ranges for more than 10 million years have caused an orographic shadow effect that has prevented the passage of humid winds from the Pacific Ocean, causing the generation of dry forests that predominate in this region (Becerra, 2005; Gómez-Tuena et al., 2005). This complex part of Mexico is older than the eastern portion of the TMVB and parts of the Central-Northeastern region, and its evolution is documented by orogenic episodes of the Late Miocene mafic episodes, which began in the West-Central part of Mexico and progressed eastward (Gómez-Tuena et al., 2005). In BAL there are recorded 20 species, all of them well represented in the tropical dry forests; however, most are also found in the temperate forest surrounding the basin. As with BAL, in other parts of the country because of the orographic shadow effect caused by the emergence of these mountain ranges, the development of dry forests and arid zones was favored.

Table 9.5 Occurrence of species of *Melampodium* within floristic regions of Mexico

	Floristic regions[a]						
Species	1	2	3	4	5	6	Total
M. americanum		X	X	X	X	X	5
M. appendiculatum			X				1
M. argophyllum				X			1
M. aureum				X	X	X	3
M. bibracteatum			X	X	X		3
M. cinereum			X	X			2
M. cupulatum	X		X				2
M. dicoelocarpum			X		X		2
M. diffusum					X	X	2
M. divaricatum		X	X	X	X	X	5
M. glabribracteatum						X	1
M. glabrum				X	X		2
M. gracile		X	X	X	X	X	5
M. leucanthum			X	X			2
M. linearilobum			X		X	X	3
M. longicorne			X				1
M. longifolium				X	X	X	3
M. longipes			X		X	X	3
M. longipilum				X	X	X	3
M. mayfieldii					X		1
M. microcephalum			X	X	X	X	4
M. mimulifolium			X		X	X	3
M. moctezumum			X				1
M. montanum				X	X	X	3
M. nayaritense			X				1
M. northingtonii						X	1
M. nutans					X	X	2
M. paniculatum					X	X	2
M. perfoliatum			X	X	X	X	4
M. pilosum					X		1
M. pringlei						X	1
M. repens			X	X	X	X	4
M. rosei			X		X		2
M. sericeum			X	X	X	X	4
M. sinaloense			X		X		2
M. sinuatum	X						1
M. strigosum			X	X	X	X	4
M. tenellum			X		X		2
M. tepicense			X		X		2
Totals	2	3	24	17	27	21	

After Villaseñor and Ortiz (2025)

[a] *(1) Baja California peninsula* (Baja California, Baja California Sur); *(2) Yucatán peninsula* (Campeche, Quintana Roo, Tabasco, Yucatán); *(3) Northwestern* (Chihuahua, Durango, Nayarit, Sinaloa, Sonora); *(4) Central-Northeastern* (Aguascalientes, Coahuila, Guanajuato, Hidalgo, Nuevo León, Querétaro, San Luis Potosí, Tamaulipas, Zacatecas); *(5) West-Central* (Colima, Guerrero, Jalisco, México, Michoacán, Morelos, Tlaxcala); *(6) Southeastern* (Chiapas, Oaxaca, Puebla, Veracruz)

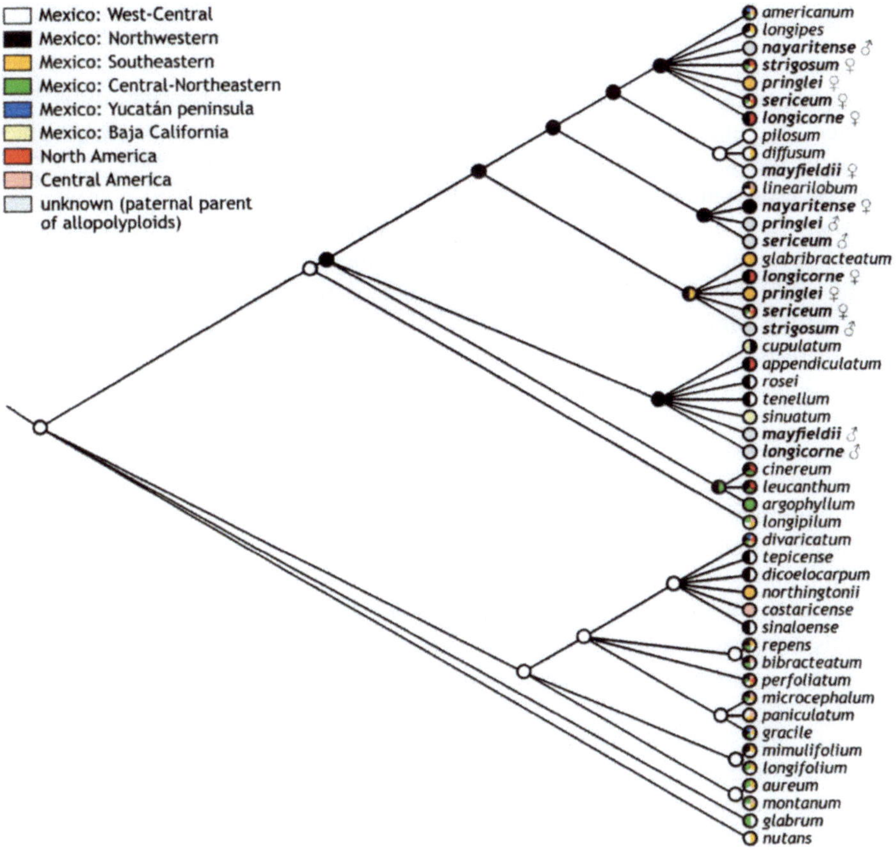

Fig. 9.4 Ancestral area reconstruction in *Melampodium*. Species in bold are allopolyploids represented by their subgenomes; male and female symbols refer to parentage

In correlation with these major geological events, the greatest richness of *Melampodium* species is found in the mountainous parts of this West-Central region (extreme south of SMOC, SMS, and TMVB), as well as in its tropical forests (BAL) (Table 9.5). In addition, the earliest diversified species, belonging to sections *Glabrata*, *Nutantia*, and *Rhizomaria*, are concentrated along this West-Central region. Several of these species show dispersal to other regions, mainly using the connection between mountain ranges, for example, *M. glabrum* is also found in parts of the Central-Northeastern region; likewise, *M. nutans* distributes also toward the Southeastern region.

Phylogeography of the White-Rayed Complex

Phylogeography is an approach to understanding how genetic variability is distributed over the landscape with particular import for inferring biogeographic trends at specific, infraspecific, and populational levels (Avise, 2000, 2008). A suitable group in which to assess such trends is the white-rayed complex of *Melampodium*. Decades earlier, based on morphology and locality data, Stuessy (1971b) investigated evolutionary and biogeographic relationships among the taxa of this complex (Fig. 9.5). Because it occurs on the northernmost edge of the range of the genus, and because it is the only group that possesses white ray florets, there is little doubt that the group is one of the most highly derived in the genus. Stuessy (1971b) proposed that the white-rayed complex originated from an ancestor similar morphologically and chromosomally to *M. americanum*. This species is perennial, as also are all members of the white-rayed complex, and it is diploid ($n = 10$), as with *M. cinereum* and *M. leucanthum*, although these each also harbor tetraploid plants within their ranges. The chromosome number of *M. argophyllum*, treated initially as a variety of *M. leucanthum* (Stuessy, 1969b), was at that time unknown, but recent cytological studies have shown it to be hexaploid with $n = 30$ (Weiss-Schneeweiss et al., 2009). *Melampodium americanum* also extends northward into Tamaulipas, in the northeastern region of Mexico, which seemed like a reasonable biogeographic connection for the origin of the white-rayed complex.

With the assumption that the white-rayed complex originated from an ancestor of *M. americanum*, an initial hypothesis for the diversification of the group was presented (Stuessy, 1971b; Fig. 9.6). The idea was that an ancestral complex developed in southeastern Coahuila and northwestern Nuevo León. One line of migration and eventual speciation occurred toward the northeast into the sandy plains on both sides of the Río Grande, extending northward, and finally stopping at the Edward's Plateau. This lineage became *M. cinereum*, and three varieties diverged, var. *ramosissimum* toward the east into northern Tamaulipas, another lineage northward into the Mesquite grasslands that became var. *hirtellum*, and the third moving northeast into southern Texas that became var. *cinereum*. It was believed that *M. argophyllum*, treated at that time as a variety of *M. leucanthum*, was the transitional taxon that colonized the low mountains of Coahuila that joined with those of the Big Bend area of Texas. From this migration and speciation developed the initial populations of *M. leucanthum*. It then dispersed northward into the Texas Panhandle and adjacent regions, eastward across central Texas, stopping at the Edward's Plateau, and westward into the calciferous soils of New Mexico and Arizona.

An important advance in understanding the origin and evolution of the white-rayed complex resulted from molecular phylogenetic studies by Blöch et al. (2009). Nuclear ribosomal ITS and chloroplast *mat*K sequences revealed that Ser. *Leucantha*

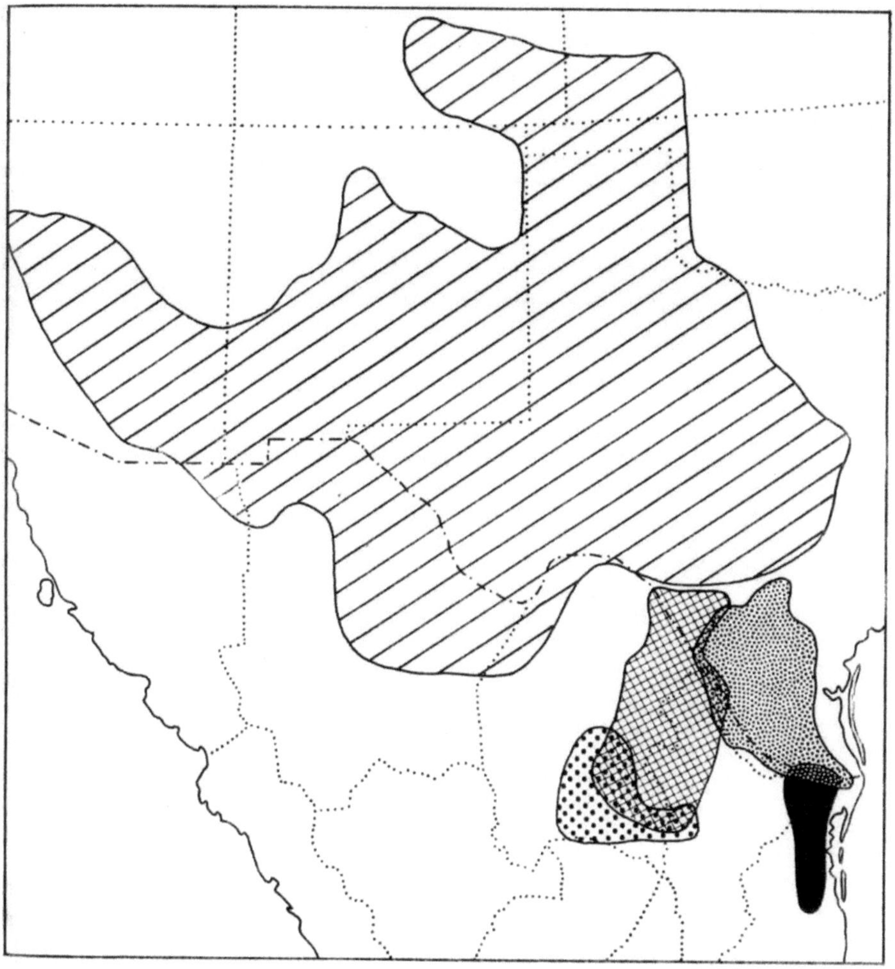

Fig. 9.5 Generalized distributions of taxa in the white-rayed complex of *Melampodium* (Sect. *Melampodium* Ser. *Leucantha*). *M. leucanthum* (diagonal lines); *M. argophyllum* (dots); *M. cinereum* var. *hirtellum* (hatched); *M. cinereum* var. *cinereum* (stippled); *M. cinereum* var. *ramosissimum* (shaded). (From Stuessy, 1971b: 179)

ties closest to Ser. *Cupulata* (see Fig. 6.4). This is a significant result because the species in the latter series are concentrated in northwestern Mexico. Two of them, *M. longicorne* and *M. appendiculatum*, also are known from southeastern Arizona. A morphological character serves to unite species of Ser. *Cupulata* and those of Ser. *Leucantha*: the cupulate series of outer phyllaries (as the name of the former series suggests). The degree of connation of the outer phyllaries into a shallow cup is

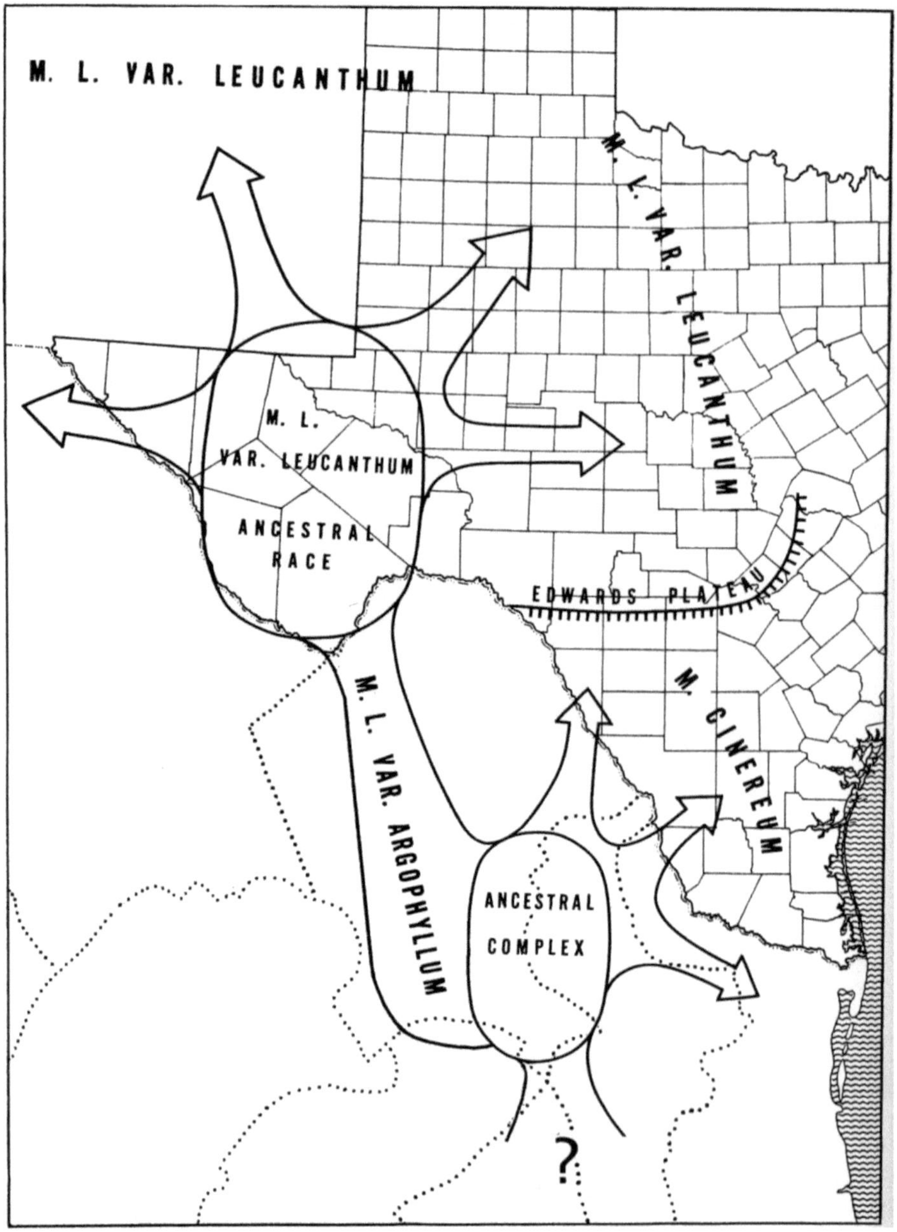

Fig. 9.6 Initial biogeographic hypothesis for the origin and spread of the white-rayed complex of *Melampodium*. (From Stuessy, 1971b: 188)

essential for distinguishing *M. leucanthum* from *M. cinereum*. This is, in fact, the very best diagnostic feature for their discrimination. It might be anticipated, therefore, that the ancestor of the white-rayed complex might well also have had a cupulate outer series of phyllaries, and this is most clearly seen among the species of Ser. *Cupulata*. *Melampodium appendiculatum*, one of the most northern-reaching species of this series, has the outer phyllaries fused nearly ¾ of their length (Stuessy, 1972), as in *M. leucanthum*. These new molecular studies reject the hypothesis of the white-rayed complex having originated on the eastern side of Mexico as shown in Fig. 9.6. Rather, the analyses strongly suggest a new hypothesis that Ser. *Leucantha* originated in Northwestern Mexico and/or the southwestern portion of the United States, perhaps in what is now Arizona. If this were the case, then migration would have taken place from west to east, rather than the reverse as initially hypothesized.

A test of the new hypothesis for the evolutionary origin and biogeography of the white-rayed complex was undertaken by Rebernig et al., (2010a, b). Her data-gathering approach relied on cpDNA sequences and amplified fragment length polymorphisms (AFLPs) to assess genetic variation within and among populations across the entire white-rayed complex. Even more important was the realization that the evolution of the entire group would have been impacted by Pleistocene glaciation and changes in the environment of the southwestern United States and adjacent regions.

Silica gel samples from 92 populations (377 individuals) of *M. leucanthum* were collected throughout its entire range. The specific questions focused on determining the spatial patterns resulting from post-Pleistocene aridification and subsequent migration pathways, plus investigating the origin and evolution of the tetraploids. It was important to identify refugial areas in this species and to compare with paleoclimatic and phylogeographic data in other, mostly animal, taxa.

Results from the AFLPs showed a division of populations into three basic groups (Fig. 9.7). The western group extends from the far western edge of distribution of the species in Arizona to eastern New Mexico, the central group occupies western Texas and adjacent New Mexico, plus extensions northward, and the eastern group occurs mostly east of the Pecos River and up into the Texas Panhandle, finishing eastward on the Edward's Plateau. Results from plastid haplotype analyses (Fig. 9.8) correspond with those from AFLPs, again revealing three groups, western, central, and eastern.

The existence of three genetic groups within the range of *M. leucanthum* suggests that they originated from three separate Pleistocene refugia. The western group, to the west of the Continental Divide, corresponds to the lower Colorado basin refugium identified previously by other workers (Hunter et al., 2001; Jaeger et al., 2005; Castoe et al., 2007; Fehlberg & Ranker, 2009). Ecological niche models (Fig. 9.9) support another refugium in central to southern Chihuahuan Desert, also previously identified by other workers (Hunter et al., 2001; Riddle & Hafner, 2006; Castoe et al., 2007). A final, less clear refugium might have been centered around the Tamaulipan Plains (Castoe et al., 2007; Rebernig et al., 2010a).

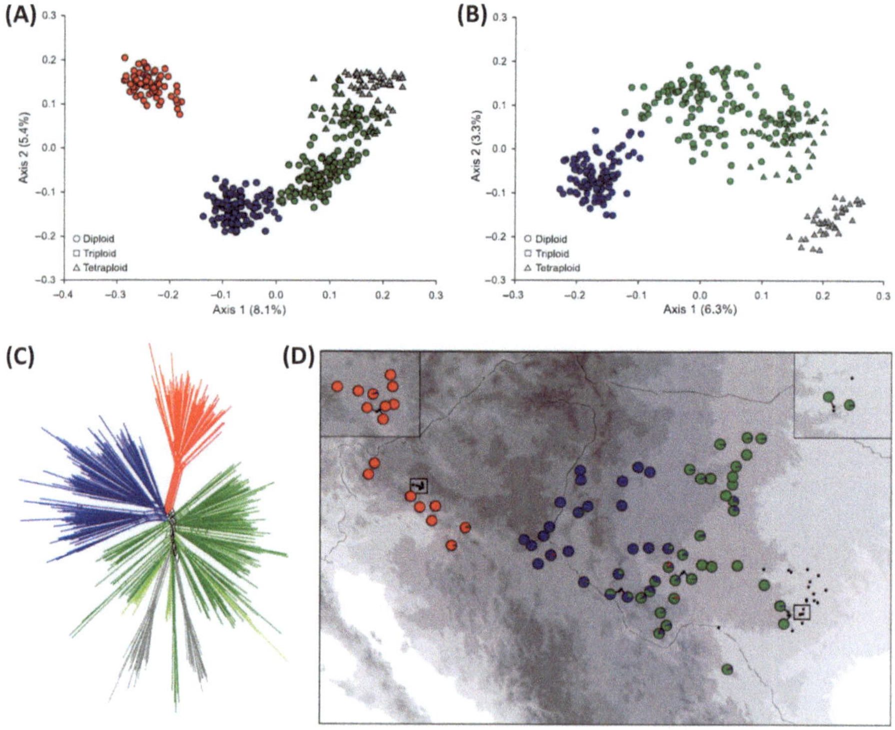

Fig. 9.7 Genetic structure of *Melampodium leucanthum* inferred from AFLP data. (**A**) Principal Coordinate Analysis (PCA) of the whole data set, and (**B**) excluding the western populations; (**C**) neighbor-net analysis of the whole data set; (**D**) STRUCTURE analysis of a data set excluding the tetraploid populations and those of unknown ploidy level (indicated by small black dots). Colors refer to the western (red), central (blue), eastern (green) and tetraploid group (gray). In (**C**) tetraploids are indicated by thick lines (and lighter green in case of the eastern group). (From Rebernig et al., 2010a: 3432)

Paleoclimatic data have suggested that Holocene aridification in the southwestern United States proceeded from west to east (Holmgren et al., 2007), but the analyses do not confirm that ancestral populations of *M. leucanthum* followed this route at the end of the Pleistocene. Instead, it is suggested that independent range shifts took place out of several refugia. The Continental Divide is a substantial barrier to a west-east migration post-Pleistocene, and in fact, the strong separation between western and central groups in *M. leucanthum* may be accounted for by this barrier. The genetic overlap of some populations in central and eastern parts of the range suggests that there might have been intermixing as the migrations continued northward and eastward from the Chihuahuan Desert.

Rebernig et al. (2010b) investigated similar populational relationships within *M. cinereum*. The difference in this species, in contrast to *M. leucanthum*, is that it shows morphological divergence over its range, such that it has been divided taxonomically into three nearly allopatric varieties, var. *cinereum*, var. *hirtellum*, and

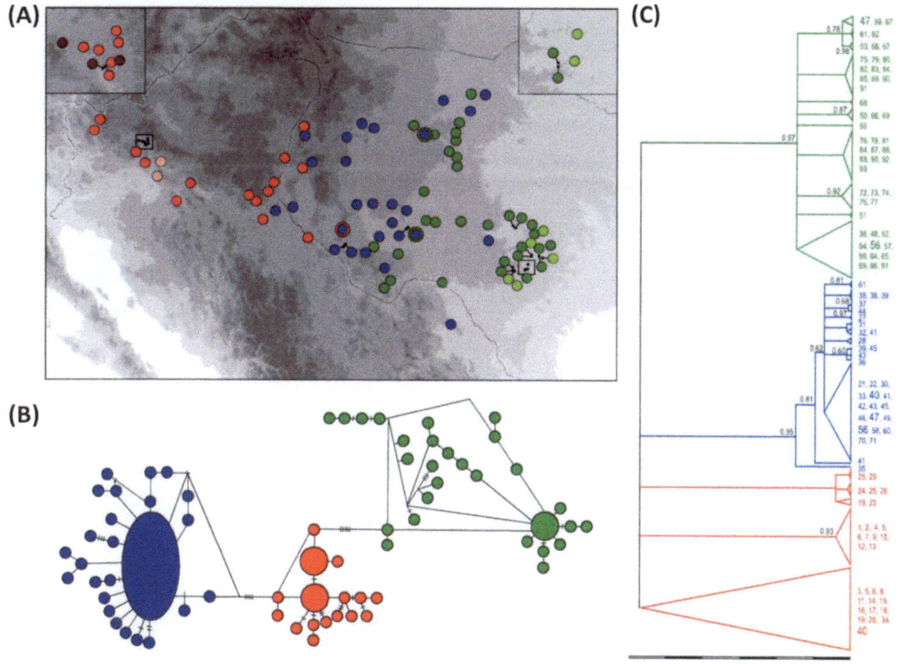

Fig. 9.8 Genetic structure of *Melampodium leucanthum* inferred from plastid sequences. (**A**) Spatial analysis of molecular variance (SAMOVA); (**B**) statistical parsimony network (unsampled haplotypes indicated by ticks); (**C**) relaxed clock Bayesian analysis with a demographic model of constant population size (node heights correspond to median ages; clades have posterior probabilities of 0.99 or more unless noted otherwise, their size being proportional to the height of the triangles; terminals are specific populations; scale bar with increments of 0.25 million years). The three main groups are indicated by colors. In (**A**) additionally the circumscription of the groups with $K = 6$, which has the highest F_{CT} value, is indicated by different shades of red (western group) and green (eastern group). Populations harboring haplotypes belonging to two different groups are highlighted in (**A**) by an outline color corresponding to the second group involved or in (**C**) by a larger font of the population numbers. (From Rebernig et al., 2010a: 3434)

var. *ramosissimum* (Fig. 9.5). AFLPs and cp haplotypes were used to infer relationships among 25 populations (Fig. 9.10; involving 113 individuals with AFLP, and 77 with cpDNA). The objectives were to compare patterns of genetic variation with the established taxonomic hypothesis and to infer biogeographic differentiation, especially involving effects of post-Pleistocene migration.

The results from AFLPs revealed a clear genetic differentiation among the varieties within *M. cinereum* (Fig. 9.11). The genetic core consists of var. *cinereum* with both $2x$ and $4x$ forming populational subsets. The western populations of var. *hirtellum* are distinct from those of the eastern group, although both fall within the generalized distribution of the variety (Fig. 9.5). Variety *ramosissimum*, occurring in the northern Tamaulipan plain region, is different from the other varieties. Interestingly, data from the cp haplotypes show no clear differentiation among the varieties (Fig. 9.12).

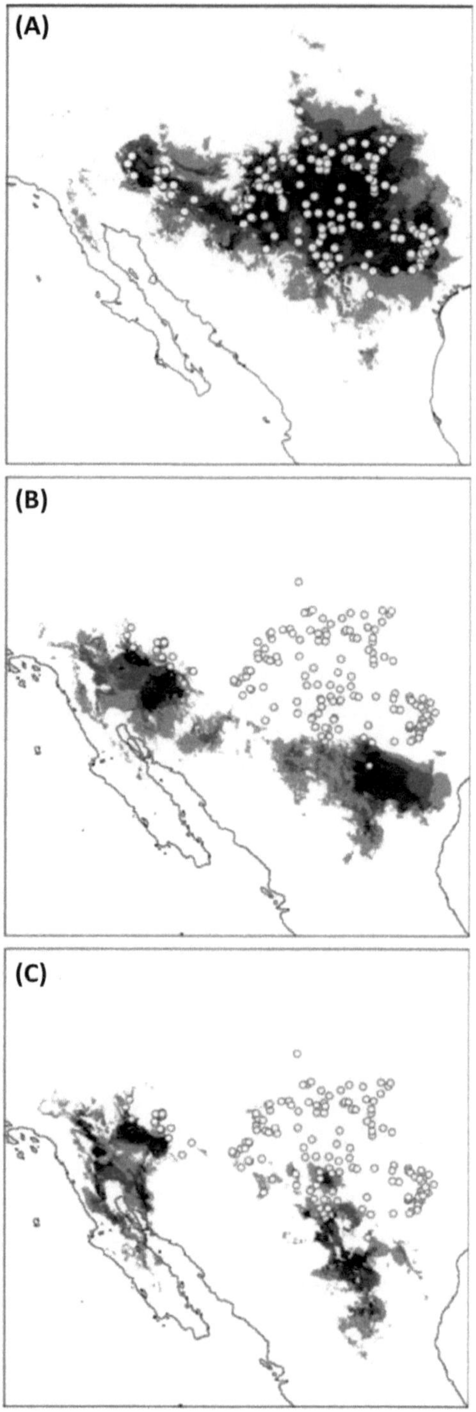

Fig. 9.9 Ecological niche modeling of *Melampodium leucanthum* for (**A**) the present and (**B**, **C**) the last glacial maximum (c. 21,000 years BP). The paleodistributions were modelled using the CCSME (in **B**) and the MIROC3.2 (in **C**) climatic models. White dots represent the 164 localities of *M. leucanthum* used for ecological niche modelling. Darker shading indicates higher climatic suitability. In (**B**) and (**C**) the coastline is that of the last glacial maximum. (From Rebernig et al., 2010a: 3437)

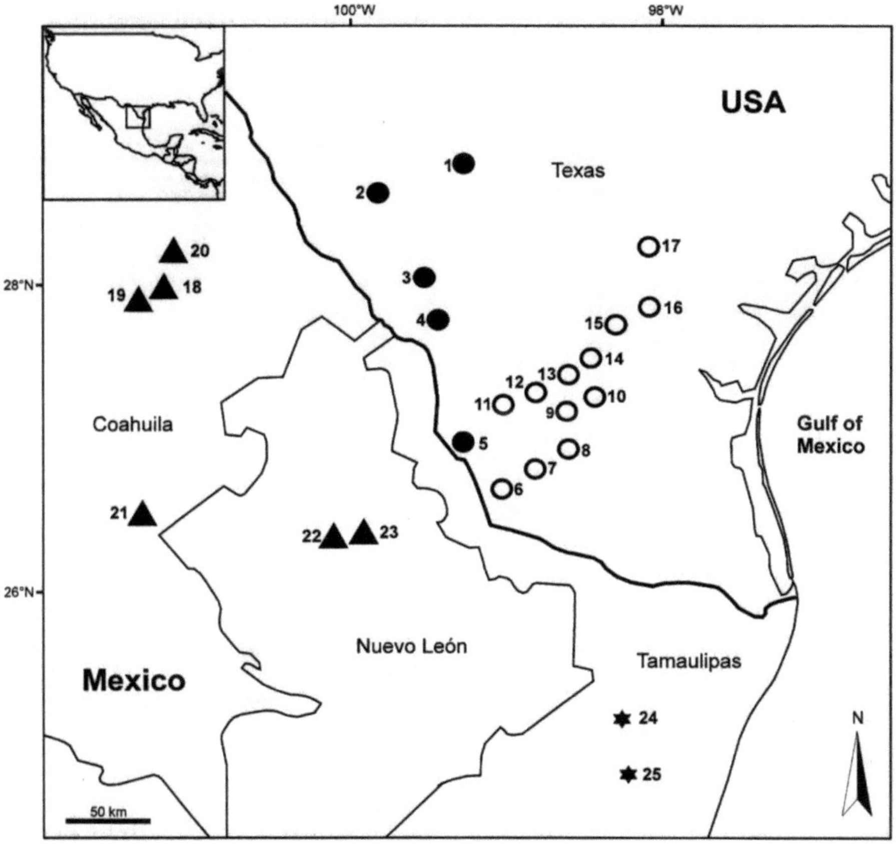

Fig. 9.10 Distribution map of populations analyzed for AFLP and cp haplotypes in *Melampodium cinereum* var. *cinereum* (closed circles, diploids; open circles, tetraploids), var. *hirtellum* (triangles) and var. *ramosissimum* (stars). (From Rebernig et al., 2010b: 596)

Environmental impact from Pleistocene glaciation in this area of the southwestern United States and adjacent Mexico was presumed similar to that which involved more western regions and *M. leucanthum*. Lowering of temperature and increased moisture during the glacial period would have pushed previously adapted taxa to arid zones further south, into refugia such as in the southern Chihuahuan Desert. Post Pleistocene migration would have occurred northward into the present patterns of distribution, stopping at the Edward's Plateau, and accompanied by morphological differentiation from geographic isolation and adaptations into distinct microenvironments. It is presumed that the basic separation into the three lineages of the white-rayed complex that led to *M. leucanthum*, *M. cinereum*, and *M. argophyllum* had already become established before migration southward into the Pleistocene refugia. Northward movement after the Ice Age may have led to some intermixing of differentiating varietal lineages within *M. cinereum*, which might account for the lack of distinct cp haplotypes seen today.

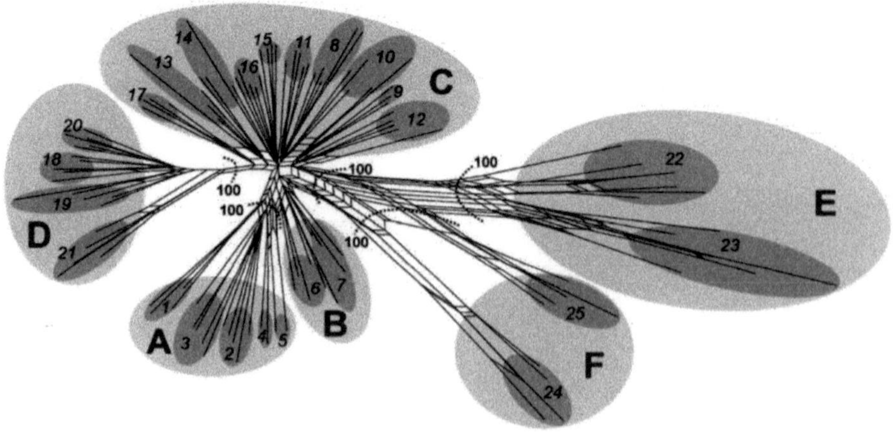

Fig. 9.11 Neighbor net for the AFLP data using Nei-Li distances among populations and taxa of *M. cinereum*. Circumscription of the groups: A = diploid var. *cinereum*; B, C = tetraploid var. *cinereum*; D = western var. *hirtellum*; E = eastern var. *hirtellum*; F = var. *ramosissimum*. Numbers in italics refer to populations. Numbers in normal font are bootstrap values from a neighbor joining analysis. (From Rebernig et al., 2010b: 600)

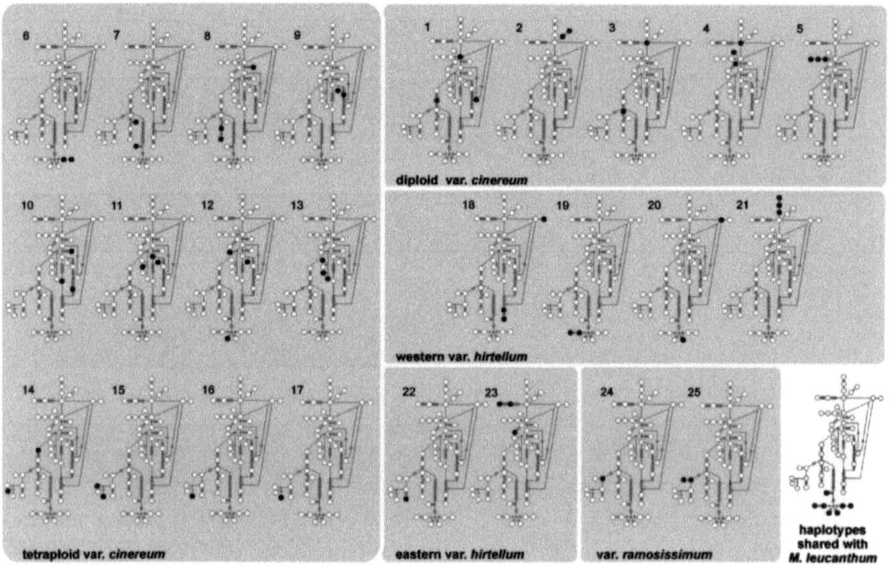

Fig. 9.12 Populations (1–25) of *M. cinereum* analyzed for cp haplotype networks. Sampled haplotypes are indicated by circles (black ones indicating those found in the respective populations), unsampled ones by ticks. The last and unnumbered network shows the haplotypes that are shared with *M. leucanthum* (marked in black). (From Rebernig et al., 2010b: 602)

Adventive Species

Several *Melampodium* species have become successful invasives in other parts of the world. The most conspicuous in this regard is *M. divaricatum*. This species has several attributes that likely have allowed it to become a successful global traveler. The natural range of the species within Mexico is very broad (Fig. 4.2A), which attests to a wide physiological tolerance. It is an annual species that prefers disturbed soils, especially in areas impacted by people. The stems root easily at the nodes, and when stems bend down and touch the ground, adventitious roots will form, which allows asexual reproduction via fragmentation if detached from the parental plant. Furthermore, breeding studies (Stuessy, unpublished) have shown the species to be self-fertilizing, which allows the establishment of new populations from a single invading individual. These features have led to *M. divaricatum* colonizing other countries outside its natural range, beginning with southeastern Brazil. Other localities include the Philippine Islands, Cuba, Virgin Islands, Burma (Myanmar), and Puerto Rico, at minimum. This species has also been brought into the horticultural trade (see Chap. 2), which encourages broader distribution through escape from gardens. Other species have also appeared previously in odd localities, such as *M. perfoliatum* in Cuba and Los Angeles, California (Stuessy, 1972).

Biogeography of Related Genera

The genera closest to *Melampodium* are *Acanthospermum* and *Lecocarpus*, both also in Melampodiinae of Millerieae. Molecular phylogenetic investigations (Blöch et al., 2009) using ITS and *mat*K have provided a perspective on the relationships of *Melampodium* to these two genera that helps inform their biogeography. In both analyses, the smaller two genera unite in a clade not containing any species of *Melampodium*. With *mat*K they form, together with *M. glabrum*, a clade that is sister group to the remainder of *Melampodium* (Fig. 6.5), and with ITS the clade of *Acanthospermum* and *Lecocarpus* is placed as sister group to Sect. *Melampodium* (Fig. 6.4), although with limited support. With chloroplast data, *Lecocarpus* and *Acanthospermum* samples maintain generic boundaries, but with ITS, *A. australe* instead groups with *Lecocarpus*, yet with limited support only. From an evolutionary taxonomic perspective, the senior author (Stuessy) prefers to maintain the three genera as distinct. In this context, it seems probable that *Acanthospermum* evolved from out of *Melampodium* somewhere in southern Mexico, perhaps in Chiapas. The former genus migrated southward throughout Central America and then spread out into the Andean region, subsequently dispersing to other countries of the continent. From a source in Ecuador, perhaps resembling *A. microcarpum*, dispersal by long-distance took place to the Galápagos Islands, where over time, divergence in isolation resulted in the evolution of three species on different islands of the archipelago.

Chapter 10
Systematics

Investigations of the systematics of a group of plants are of great significance, because it is through these studies that predictive classifications can be constructed and evolutionary processes can be inferred. Since the original description of *Melampodium* over 270 years ago, much effort has been expended to (1) determine the limits of the genus, (2) provide a structure of species into sections and series, and (3) understand the boundaries for all species. These aspects of classification have been dealt with to some extent in other chapters, involving discussions of evolution and biogeography, but the present chapter examines in more detail the history of concepts of classification at different levels of the taxonomic hierarchy.

Original Description of the Genus

As with many genera of flowering plants, *Melampodium* was first described by Carl Linnaeus in his *Species Plantarum* (1753), and the genus contained the single species *M. americanum* (Fig. 10.1A). More details of the morphology of the genus were provided in his *Genera Plantarum* of 1754 (Fig. 10.1B). Linnaeus makes reference to his previous publication, *Hortus Cliffortianus* (1737a), and on p. 425 of that work (Fig. 10.2A) is a long description and additional information about the species and original specimen. The description is detailed, including a final comment that the fruit (and enclosing phyllary with hood) resembles "*pedem caprae,*" the foot of a goat (which in some species it indeed does; see Figs. 3.3D and 3.4).

He also commented that the species grows in "Vera Cruce" (Veracruz, Mexico), and that it was communicated by Ph. [Philip] Miller (1691–1771), well known as the author of *The Gardner's Dictionary* (1731). After the listed polynomial synonym, "Caltha americana, foliis laciniatis, flore luteo," Linnaeus also wrote "Houst. mss." This refers to William Houstoun (1695–1733), a British surgeon who had studied in Leiden with Prof. Herman Boerhaave, and who sailed (1729–1733) to

T. F. Stuessy et al., *Systematics, Ecology, and Chromosomal Evolution of* Melampodium *(Asteraceae)*, https://doi.org/10.1007/978-3-031-81995-7_10

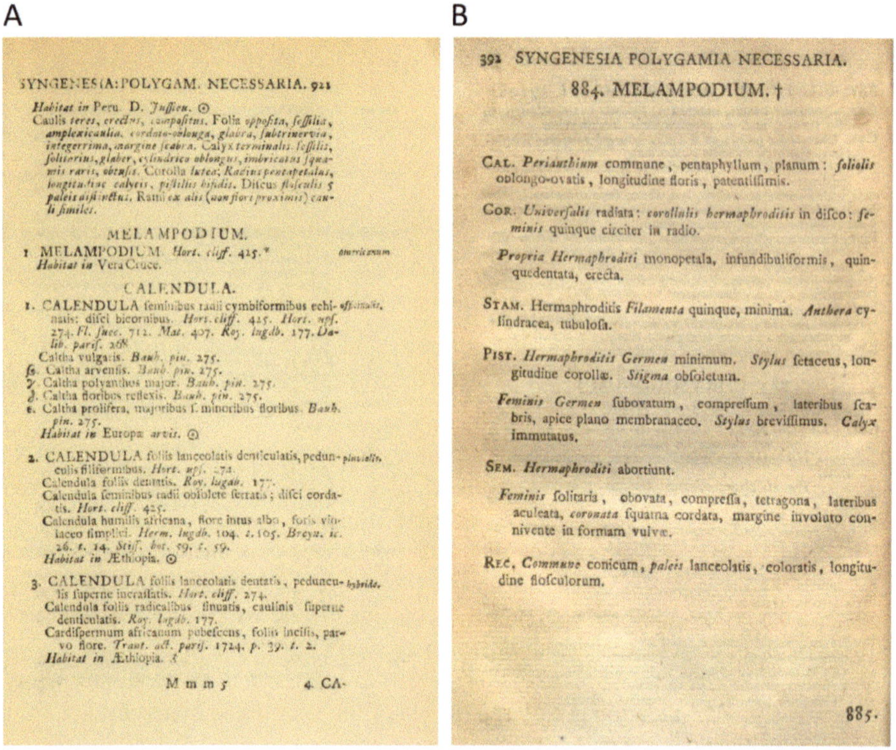

SYNGENESIA: POLYGAM. NECESSARIA. 921

Habitat in Peru. D. *Juſſieu.* ☉

Caulis *teres, erectus, compoſitus.* Folia *oppoſita, ſeſſilia, amplexicaulia, cordato-oblonga, glabra, ſubtrinervia, integerrima, margine ſcabra.* Calyx *terminalis, ſeſſilis, ſolitarius, glaber, cylindrico oblongus, imbricatus ſquamis raris, obtuſis.* Corolla *lutea; Radius pentapetalus, longitudine calycis, piſtillis bifidis.* Diſcus *floſculis 5 paleis diſtinctus.* Rami *ex alis (non flori proximis) cauli ſimiles.*

MELAMPODIUM.

1. MELAMPODIUM. *Hort. cliff.* 425.* — *americanum*
Habitat in Vera Cruce.

CALENDULA.

1. CALENDULA ſeminibus radii cymbiformibus echinatis: diſci bicornibus. *Hort. cliff.* 425. *Hort. upſ.* 274. *Fl. ſuec.* 712. *Mat.* 407. *Roy. lugdb.* 177. *Dalib. pariſ.* 268. — *officinalis.*
Caltha vulgaris. *Bauh. pin.* 275.
β. Caltha arvenſis. *Bauh. pin.* 275.
γ. Caltha polyanthos major. *Bauh. pin.* 275.
δ. Caltha floribus reflexis. *Bauh. pin.* 275.
ε. Caltha prolifera, majoribus ſ. minoribus floribus. *Bauh. pin.* 275.
Habitat in Europæ *arvis.* ☉

2. CALENDULA foliis lanceolatis denticulatis, pedunculis filiformibus. *Hort. upſ.* 274. — *pluvialis.*
Calendula foliis dentatis. *Roy. lugdb.* 177.
Calendula ſeminibus radii obſolete ſerratis; diſci cordatis. *Hort. cliff.* 425.
Calendula humilis africana, flore intus albo, foris violaceo ſimplici. *Herm. lugdb.* 104. *t.* 105. *Breyn. ic.* 26. *t.* 14. *Stiſſ. bot.* 59. *t.* 59.
Habitat in Æthiopia. ☉

3. CALENDULA foliis lanceolatis dentatis, pedunculis ſuperne incraſſatis. *Hort. cliff.* 274. — *hybrida.*
Calendula foliis radicalibus ſinuatis, caulinis ſuperne denticulatis. *Roy. lugdb.* 177.
Cardiſpermum africanum pubeſcens, foliis inciſis, parvo flore. *Trant. act. pariſ.* 1724. *p.* 39. *t.* 2.
Habitat in Æthiopia.

Mmm 5 — 4. CA-

392 SYNGENESIA POLYGAMIA NECESSARIA.

884. MELAMPODIUM. †

CAL. *Perianthium* commune, pentaphyllum, planum: *foliolis* oblongo-ovatis, longitudine floris, patentiſſimis.

COR. *Univerſalis* radiata: *corollulis hermaphroditis* in diſco: *femininis* quinque circiter in radio.

Propria Hermaphroditi monopetala, infundibuliformis, quinquedentata, erecta.

STAM. Hermaphroditis *Filamenta* quinque, minima. *Anthera* cylindracea, tubuloſa.

PIST. *Hermaphroditis Germen* minimum. *Stylus* ſetaceus, longitudine corollæ. *Stigma* obſoletum.

Femininis Germen ſubovatum, compreſſum, lateribus ſcabris, apice plano membranaceo. *Stylus* breviſſimus. *Calyx* immutatus.

SEM. *Hermaphroditi* abortiunt.

Femininis ſolitaria, obovata, compreſſa, tetragona, lateribus aculeata, *coronata* ſquama cordata, margine involuto connivente in formam vulvæ.

REC. *Commune* conicum, *paleis* lanceolatis, coloratis, longitudine floſculorum.

885.

Fig. 10.1 Original description (**A**) of *Melampodium americanum* from the *Species Plantarum* of Linnaeus (1753: 921), and (**B**) accompanying generic description in the fifth edition of his *Genera Plantarum* (1754: 392)

Mexico and the West Indies with the South Sea Company (Dandy, 1958; Stafleu & Cowan, 1979). Houstoun collected *M. americanum* near Veracruz, where it is native (see Fig. 4.3B). After Houstoun died in 1733, his collections, papers, and drawings were bequeathed to Miller (Stafleu & Cowan, 1979).

Of particular relevance is the information about *Melampodium* contained in the manuscript of Houstoun (and communicated by Miller) that Linnaeus cited in the *Hortus Cliffortianus*. There are two literature sources for this information. The first comes from materials by Houstoun given by Philip Miller to Linnaeus, when the latter visited England in 1736 while he was in the employ of Clifford (Stearn, 1960). Linnaeus called on Philip Miller at the Chelsea Physic Garden, but the meeting initially was cold. After a time, things warmed up, and Miller finally gave Linnaeus seeds and specimens to take back to Clifford, which had been a request his benefactor had made in exchange for paying for Linnaeus's trip abroad. Included with these materials was the original specimen (lectotype; Jarvis, 2007) of *Melampodium americanum* (Fig. 10.2B). According to Stafleu and Cowan (1979), Miller also gave Linnaeus unpublished notes and engravings of Houstoun that were later bound together under the title (written in Linnaeus's hand; Fig. 10.3A) of *Houstonia*

A B

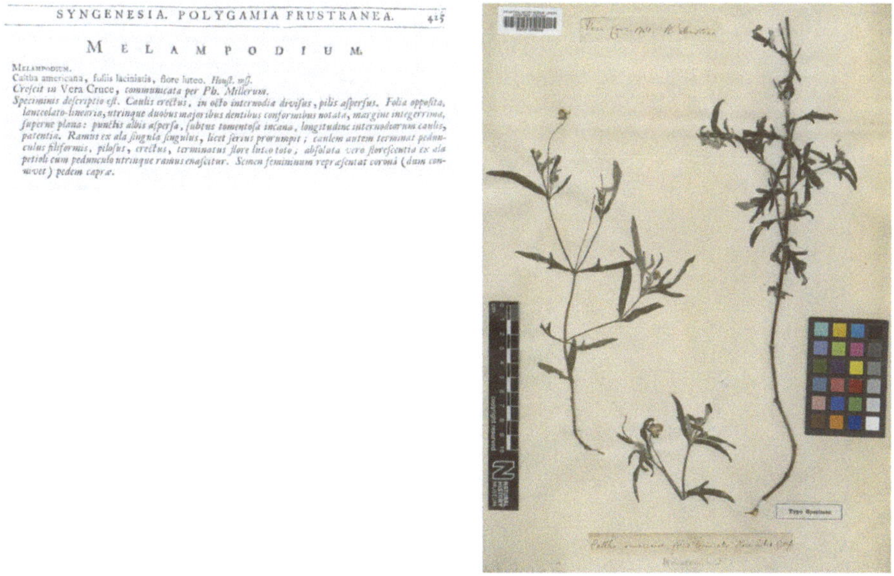

SYNGENESIA. POLYGAMIA FRUSTRANEA. 415

M E L A M P O D I U M.

Melampodium.
Caltha americana, foliis laciniatis, flore luteo. *Houst. mss.*
Crescit in Vera Cruce, *communicata per Ph. Millerum.*
Speciminis descriptio est. Caulis erectus, in octo internodia divisus, pilis aspersus. Folia opposita, lanceolato-linearia, utrinque duobus majoribus dentibus conformibus notata, margine integerrima, superne plana: punctis albis aspersa, subtus tomentosa incana, longitudine internodiorum caulis, patentia. Ramus ex ala singula singulus, licet serius prorumpit; caulem autem terminat pedunculus filiformis, pilosus, erectus, terminatus flore luteo toto; absoluta vero florescentia ex ala petioli cum pedunculo utrinque ramus enascitur. Semen femininum repræsentat coronâ (dum connivet) pedem capræ.

Fig. 10.2 Description of *Melampodium* by Linnaeus (**A**) in his *Hortus Cliffortianus* (1737a); original specimen (**B**) of *Melampodium americanum* from the Clifford Herbarium now housed in the British Museum (BM)

Genera Plantarum Americanarum nova communicata per Phil. Millerum, Lond. 1736. This was, in essence, a manuscript, a set of 15 pages of plant illustrations, but without descriptions. *Melampodium americanum* is strangely not included, either with text or illustration. Clearly Linnaeus must have had a more complete copy with information on the locality of *Melampodium americanum*, or he would not have been able to cite Vera Cruce as locality in the *Hortus Cliffortianus*. It is possible that some plates (and/or descriptions) were left out (or lost) before binding. Linnaeus took materials back to Clifford's estate, the Hartekamp, and he used it for work on *Hortus Cliffortianus*. Upon Linnaeus's return to Sweden in 1738, he took the Houstoun materials with him, except for specimens that remained with Clifford. When Linnaeus died in 1778, his collections and library were sold by his widow, Sara Lisa, in 1784. They were sent to London to the purchaser, James Edward Smith (1759–1828), who subsequently used them as a nucleus around which to found with other botanists the Linnean Society of London. Upon Smith's death in 1828, the Linnaean natural history collections and library were sold to the Linnean Society, where they remain today. This is how the Houstoun manuscript made its way back to London.

The second source of information from Houstoun comes from a small publication (Fig. 10.3B) by Joseph Banks in 1781 in London entitled *Reliquiae Houstounianae: seu plantarum in America meridionali a Gulielmo Houstoun*

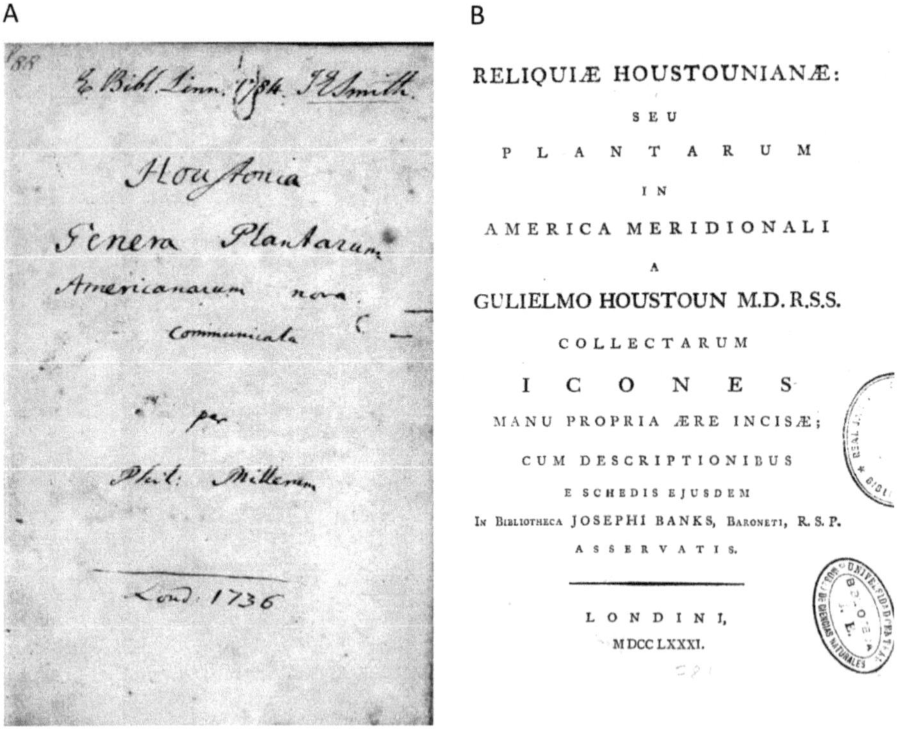

Fig. 10.3 Text and illustrations on American plants by William Houstoun given to Phillip Miller and then to Linnaeus in 1736 (**A**), later published by Joseph Banks in 1781 (**B**)

M.D.R.S.S. collectarum.... After Miller's death in 1771, his collections (still containing Houstoun materials) were sold to Joseph Banks in London, who took it upon himself to publish the small book. This work consists of 10 pages of notes and descriptions of plants plus 26 illustrations. The description of *Caltha* (*Melampodium*) *americana* (Fig. 10.4a) is lengthy and detailed, and it differs somewhat from what Linnaeus wrote in the *Hortus Cliffortianus*. It reads: "CALTHA Americana, foliis laciniatis, flore luteo (*a*)," which in a footnote at the bottom of the same page says "(*a*) Melampodium americanum *Linn.*," doubtless added by Banks. Also in this work is a nice illustration (Fig. 10.4b), Tab. 21. Information is provided on the origin of the specimen: "Nascitur prope Veram Crucem solo arenoso, et praecipue praerupto. Mense Martio 1731 florebat, et semen maturum ferebat." [loosely translated: Grows near Veracruz in sandy and especially steep places, flowering and producing mature seed in March of 1731]. One might surmise that Houstoun, when he originally gifted materials to Miller, may have given him several copies (perhaps proofs, because the images are printed, not hand drawn) for an intended book on plants of the New World based on his collections. This may be why Miller had no hesitation to give Linnaeus one of these copies during their first (and only) visit in 1736.

A

[9]

ſcuntur ſeſquiunciæ plerumque intervallo, conjugata, Origani vel Ocymi foliis ſimilia figura et magnitudine, averſa parte villoſa, et ut plurimum ſerrata. *Flores* ſummis inſident ramulis, ſolitarii, lutei, formæ ut in congeneribus, et magnitudinis in figura expreſſæ. *Semina* ſuccedunt exigua, compreſſa, nigricantia, ſed brevi et alba lanugine tecta.

Provenit locis apricis et humidioribus prope Veram Crucem. Menſe autem Martio paſſim florebat et ſemen perficiebat.

CALTHA Americana, foliis laciniatis, flore luteo (*a*). Tab. 21.

Multi ab eadem radice oriuntur *caules*, teretes, herbacei, villoſi et humi procumbentes. *Folia* conjugatim naſcuntur, ſeſquiunciam longa, unciam vero lata, in quinque lacinias plerumque diviſa, ſed interdum ludente natura integra, utrinque hirſuta, ſed præcipue averſa parte, ubi etiam ſunt incana. *Rami* ex alis foliorum oriuntur; *Flores* autem ſummis inſident ramulis. Horum forma congeneribus eſt ſimilis; color luteus; et magnitudo eadem quæ in figura exhibetur. At *calix* ſimplex eſt, quinque conſtans ſegmentis latis, mucronatis, et hirſutis. Floſculi, ut in aliis Calthæ ſpeciebus, ſunt ſteriles: ſemi-floſculis vero ſuccedit corona *ſeminum* breviuſculorum, aſpero cortice tectorum.

Naſcitur prope Veram Crucem ſolo arenoſo, et præcipue prærupto. Menſe Martio 1731 florebat, et ſemen maturum ferebat.

RICINOIDES folio ſubrotundo, ſerrato; fructu parvo, conglo-merato. Tab. 22.

Caules huic plantæ teretes, herbacei, et hirſuti, plerumque erecti, non raro tamen procumbentes. *Ramuli* nunc ſolitarii, nunc terni vel quaterni ſimul ex caulibus oriuntur. *Folia* ſubrotunda ſunt, pollicem lata, ſeſquipollicem longa, crenata, villoſa, pediculis ſemiuncialibus inſiſtentia, et incerto ordine poſita. In ſummis ramulis proveniunt

(*a*) Melampodium americanum *Linn.*

B parvæ

B

Fig. 10.4 Description (**A**) and illustration (**B**) of *Melampodium americanum* (= *Caltha americana*) from William Houstoun, published in 1781 by Joseph Banks (see Fig. 10.3B)

The original specimen (Fig. 10.2B) of *Melampodium americanum* had a different fate. When Clifford died in 1760, his estate passed to his elder son, who did not share the enthusiasm of his father for natural history. The bulk of the estate was auctioned off in 1788. Sir Joseph Banks purchased the herbarium in 1791, and it was shipped to Britain for his large and well-known collections at Soho Square in London. Upon Banks's death in 1820, his herbarium was willed to his librarian and assistant, Robert Brown (1773–1858), who had been working for Banks since 1810. These materials were transferred in 1827 to the British Museum along with Brown, who was appointed as the first Keeper of Botany, a position he held until his death in 1858. This explains how the original specimen of *Melampodium americanum* also came to reside in London.

Generic Limits

From the initial description by Linnaeus, most of the generic complexity of *Melampodium* during the nineteenth century resulted from decisions on maintaining a broad concept of the genus or breaking it up into several genera, more or less following present sectional lines. Two different alternatives have been seriously contemplated. The first was the submerging of the genus *Unxia* from South America into *Melampodium* by Bentham and Hooker (1873) and Baker (1884). In this small genus of two recognized species, the inner phyllaries partially and loosely enclose the ray achenes but do not fuse with them, which vitiates inclusion into *Melampodium*. The proper relationships of *Unxia* seem to be with *Smallanthus* (*Polymnia*; Stuessy, 1969a), somewhat distant from *Melampodium* within the Heliantheae alliance.

The second alternative to generic limits in *Melampodium* was incorporating the closest relatives, *Acanthospermum* and *Lecocarpus*, both of which also have inner phyllaries that enclose and fuse with the ray achenes. In *Acanthospermum* the bracts are often prickly, even hooked, and in *Lecocarpus* there are several long prickles or wing-like structures toward the apex of the fruit. In his monumental treatment of the Compositae, Bentham (1873: 434) offered the opinion that "The east tropical *Acanthospermum* and the Galapagian *Lecocarpus* might be included in the widely spread *Melampodium*." He did not formally act on this suggestion, however, although Baillon (1882) later treated the included genera as sections within *Melampodium*. Despite the unifying feature of inner phyllaries enclosing and fused with the fruits, the three genera are distinct morphologically, and their geographical centers are also different, with *Melampodium* mostly in Mexico, *Acanthospermum* more in Central America and northwestern South America (with six species; Stuessy, 1970b), and with three to four species of *Lecocarpus* (Eliasson, 1971; Adsersen, 1980; Sønderberg Brok & Adsersen, 2007; Tye & Jaramillo, 2022) endemic to the Galápagos Islands. Because of the importance of these two genera as being the closest relatives of *Melampodium*, more detail on their taxonomy is provided below.

The first revision of *Acanthospermum* appeared in 1921 by S. F. Blake, in which he treated eight species distributed into three taxonomic sections: Sect. *Lecocarpopsis* (with the single species *A. lecocarpoides* from the Galápagos Islands), Sect. *Ceratochlaena* (with five species, *A. humile*, *A. hispidum*, *A. donii*, *A. simile*, and *A. microcarpum*, the latter also from the same Islands), and Sect. *Xanthioides* (with *A. consobrinum* and *A. australe*). The non-Galápagos species were distributed in South America, and *A. australe*, likely due to the hooked fruits that aid in dispersal, has also become a weed in the eastern and southern United States, plus throughout South America and in the Hawaiian Islands and India (Blake, 1921). In 1922 Blake added two new species to the genus in Sect. *Lecocarpopsis*, *A. brachyceratum* and *A. leptolobum*, both from the Galápagos Islands, which brought the number of species of *Acanthospermum* in the archipelago to four and the total number to ten.

Stuessy (1970b) reviewed *Acanthospermum* and made several taxonomic changes. Most importantly, *A. donii*, known "probably from Ecuador" (Blake, 1921: 388), was submerged into *A. microcarpum*, originally described from the Galápagos Islands by Robinson (1902), which removed the latter as an endemic from the archipelago. This also reduced the total number of species of *Acanthospermum* to six, with a clear geographic center in mainland northwestern South America. Three species had restrictive distributions: *A. simile* from Ecuador, *A. microcarpum* from Ecuador (including the Galápagos Islands), and *A. consobrinum* from Paraguay. The other species are more widely distributed (from Blake, 1921): *A. humile* in the Caribbean and Central America; *A. hispidum* in the southeastern United States,

Central America, the Caribbean, and several countries of South America (Colombia, Venezuela, Peru, Bolivia, Brazil, Paraguay, and Argentina), plus Africa and the Hawaiian Islands; and *A. australe* with a similar, and even broader, distribution. These more broadly dispersed species have fruits with hooked awns, which may have added dispersal, especially during historical times. *A. consobrinum*, which has a limited distribution in Paraguay, also has conspicuous spines, but they are not as strongly hooked as in the other three species, perhaps limiting its dispersibility.

Lecocarpus was originally described by Decaisne (1846) from the Galápagos Islands with the single species, *L. pinnatifidus*. Cronquist and Stuessy (in Cronquist, 1970) transferred two *Acanthospermum* species of the Galápagos Islands, *A. lecocarpoides* and *A. leptolobum*, into *Lecocarpus*, as *L. lecocarpoides* and *L. leptolobus*, increasing the number of endemics of that latter genus to three (along with *L. pinnatifidus*). In the following year, Cronquist (1971a, b) submerged *A. brachyceratum* into *L. lecocarpoides*, which removed this endemic species of *Acanthospermum* from the islands; *Lecocarpus* now contained *L. lecocarpoides*, *L. leptolobus*, and *L. pinnatifidus*, all endemic to the archipelago. In parallel, Eliasson (1971) completed a new, detailed revision of *Lecocarpus*, and he also recognized the same three species. Adsersen (1980) produced another revision of the genus, this time describing a new species, *L. darwinii* from Isla San Cristóbal, and submerging *L. leptolobus* into the synonymy of *L. lecocarpoides*, still retaining three species in the genus, but of slightly different composition. Very recently, Tye and Jaramillo (2022) accepted these three species but subdivided *L. lecocarpoides* into two subspecies, subsp. *lecocarpoides* and subsp. *brachyceratus*. Because of its endemic nature in the famous islands, the genus has also received attention for wood anatomy (Carlquist & Eckhart, 1982), reproductive ecology (Philipp et al., 2004), and morphological variation within populations (Brok & Adsersen, 2007).

Panero (2007) treated *Melampodium*, *Acanthospermum*, and *Lecocarpus* as constituting subtribe Melampodiinae within the Heliantheae alliance. Baldwin (2009) followed this approach, but the subtribe was placed in the tribe Millerieae (following Panero & Funk, 2002). Previous ties to *Ichthyothere* and *Smallanthus* had been proposed (e.g., Robinson, 1981), but these genera, based on cpDNA data, now seem more distant and are placed in the subtribe Milleriinae (Panero, 2007).

The subtribal concept of Melampodiinae has a long history (Fig. 10.5) extending back to Lessing (1830), who first named and described the subtribe (as Melampodieae) to include only *Melampodium* and *Polymnia*. In 1832 he expanded the subtribe into four unranked units with many more genera: Silphieae (three genera), Millerieae (seven genera), Euxenieae (three genera), and Melampodieae (with *Melampodium* and *Centrospermum* Kunth, the latter a synonym of *Acanthospermum*; Blake, 1921). Candolle (1836) broadened the concept of the subtribe by adding representatives of the *Ambrosia* complex, i.e., the divisions Ambrosieae, Iveae, and Parthenieae. Bentham and Hooker (1873) rearranged these units into four subtribes (Melampodieae, Millerieae, Petrobieae, and Ambrosieae) within the tribe Heliantheae, encompassing 20 genera, and this structure was followed by Hoffmann (1890), with the addition of *Dugesia* and *Moonia*. Since that time several new genera were added and some were subtracted, such that 23 genera were recognized by

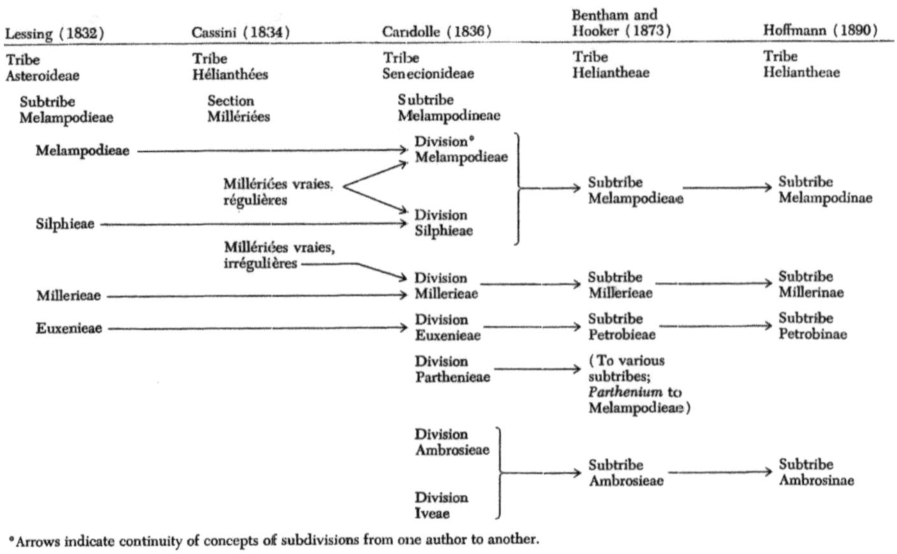

Fig. 10.5 Diagrammatic history of subtribe Melampodiinae and related groups of the Heliantheae alliance. (From Stuessy, 1973a: 67)

Stuessy (1973a) as belonging to Melampodiinae. He recommended the removal of *Parthenice* and *Parthenium* into Ambrosiinae and highlighted six informal groups that contained genera that appeared to have some semblance of relationship based on morphological and chromosomal features. Group 4 contained *Melampodium*, *Acanthospermum*, and *Lecocarpus*. In a comprehensive new rearrangement of genera within Heliantheae, Stuessy (1977) narrowed the concept of Melampodiinae to include two informal groups: (1) *Acanthospermum*, *Lecocarpus*, and *Melampodium*; and (2) *Espeletia*, *Polymnia*, *Rumfordia*, *Sigesbeckia*, *Trigonospermum*, and *Unxia*. Robinson (1981), in a restructuring of Heliantheae, treated Subtribe Melampodiinae to include *Acanthospermum*, *Ichthyothere*, *Lecocarpus*, *Melampodium*, and *Smallanthus* (= most of *Polymnia,* which was now reduced to only two species of the eastern United States, and placed in its own subtribe, Polymniinae). The relationships of Melampodiinae to other subtribes, however, were inconclusive.

Beyond *Acanthospermum* and *Lecocarpus*, the wider generic relationships of *Melampodium* are obscure. The best sampling of genera surrounding *Melampodium* in Millerieae for phylogenetic analysis was done by Rauscher (2002), who investigated five species of *Melampodium* (*M. americanum*, *M. pilosum*, *M. divaricatum*, *M. paniculatum*, and *M. gracile*) plus two species of *Acanthospermum* and one of *Lecocarpus* (Fig. 6.3). The selection of species within *Melampodium* was a good strategy because it included two in Sect. *Melampodium*, one in Sect. *Serratura*, and two in Sect. *Zarabellia*, hence also providing a small test of these sectional distinctions. The other genera included *Smallanthus*, *Ichthyothere*, plus numerous others, such as *Milleria*, *Trigonospermum*, and *Sigesbeckia*. The results (Fig. 6.3) show *Melampodium* forming a close group with *Lecocarpus* and *Acanthospermum*.

Among the species of *Melampodium*, the sectional distinctions are indicated, providing support for the classifications based on morphological and cytological data.

Subgeneric and Sectional Classification

Several subdivisions of *Melampodium* were offered during the nineteenth century. Humboldt et al. (1818), based on the inclusion of two other genera, recognized three unranked subdivisions within *Melampodium*: *Melampodia*, *Dysodia* (Rich. in Pers.) Kunth and *Alciniae* (Cav.) Kunth. Cassini (1829) took an opposite approach by recognizing *Melampodium* and *Alcina* as distinct genera, plus adding the new genus, *Zarabellia* Cass. He accepted that all three groups had the inner phyllaries enclosing and fused with the ray achenes, but that they could be distinguished by features of the disc ovaries and fruit features. *Melampodium* had conspicuous abortive ovaries of the disc florets (see Fig. 3.5A, B) and fruits crowned by a hood (Fig. 3.3D), *Zarabellia* had inconspicuous abortive disc ovaries and no hood on the fruits (although sometimes with small valve-like processes), and *Alcina* Cav. also had diminutive abortive disc ovaries and fruits without hoods or apical processes. These distinctions were very important in setting a basic structure of distinctions within the genus that we still use today. Candolle (1836), however, took a broader view of the genus and combined *Zarabellia* and *Alcina* back into *Melampodium*, but maintained them as distinct sections as *Eumelampodium*, *Zarabellia* (Cass.) DC. and *Alcina* (Cav.) DC. Baker, in his treatment (1884) of species for Martius's *Flora Brasiliensis*, recognized two subgenera: *Dysodium* (Rich. in Pers.) Kunth and *Unxia* (L.f.) Baker. The latter was originally described as a genus by Linnaeus *filius* in 1781 with the single species, *U. camphorata*, from Surinam, and subsequently transferred into *Melampodium* by Baker in 1884 as *M. camphoratum*. Baker also described the new species *M. suffruticosum* from Venezuela. Both these species were later transferred out of *Melampodium* and into the independent genus *Unxia* (Stuessy, 1969a). Baillon (1882) provided a more radical approach by combining the genera *Acanthospermum* Schrank and *Lecocarpus* Dcne. into *Melampodium*, with these genera being the basis of two sections: Sect. *Acanthospermum* (Schrank) Baillon, and Sect. *Lecocarpus* (Dcne.) Baillon.

A more complete overview of *Melampodium* was provided in the twentieth century. B. L. Robinson in 1901 produced a "synopsis," which was a step toward a revision of the genus. He submerged Sect. *Alcina* (with the sole species, *M. perfoliatum*) into Sect. *Zarabellia*, but he kept Sect. *Eumelampodium*, hence maintaining two formal taxonomic sections within the genus. Turner and King (1962) followed the two-section framework in their cytotaxonomic survey of *Melampodium*, and they did not propose any additional formal infrageneric units. They did, however, make correlations between groups of species and chromosome numbers, e.g., that species in Sect. *Melampodium* (= *Eumelampodium*) were all based chromosomally on $x = 10$. The other species of Sect. *Zarabellia* were multibasic with $x = 9$, 11, and 12. The number $n = 16$ had been counted in *Unxia camphorata*, but this no longer belonged in *Melampodium*. Turner and King did recognize some informal groups that might be worthy of formal recognition after more intensive study: (1) the small herbaceous species based on $x = 9$; (2) the rhizomatous species based on $x = 11$; and (3) annual species with large outer phyllaries and smooth achenes based on $x = 11$ and 12.

Stuessy completed a comprehensive revision of the genus (1972), in the process describing six new species (Stuessy, 1970c) and presenting a complete classification of sections and series for all species known at that time (Table 10.1). The largest Sect. *Melampodium* was maintained with 20 species, all based on $x = 10$. Five series were recognized within Sect. *Melampodium* based on constellations of morphological characters. The other sections were strongly influenced by different chromosome numbers as well as morphology, with Sect. *Zarabellia* based on $x = 9$, Sect. *Serratura* based on $x = 12$, Sect. *Rhizomaria* $x = 11$, and Sect. *Alcina* also with $x = 11$. Neither of the species in Sect. *Bibractiaria* had been counted at that time

Table 10.1 Synopsis of classification of *Melampodium* by Stuessy (1972)

I. Section *Melampodium* ($x = 10$)
- 1. Series *Melampodium*
  - 1, *M. americanum* L. 2, *M. diffusum* Cass. 3, *M. pilosum* Stuessy. 4, *M. longipes* (A.Gray) B.L.Rob. 5, *M. linearilobum* DC
- 2. Series *Leucantha* Stuessy
  - 6, *M. leucanthum* Torr. & A.Gray. 7, *M. cinereum* DC. 7a, var. *cinereum*. 7b, var. *hirtellum* Stuessy. 7c, var. *ramosissimum* (DC.) A.Gray. 8, *M. argophyllum* (A.Gray ex B.L.Rob.) S.F.Blake
- 3. Series *Sericea* Stuessy
  - 9, *M. sericeum* Lag. 10, *M. pringlei* B.L.Rob. 11, *M. strigosum* Stuessy. 12, *M. longicorne* A.Gray. 13, *M. nayaritense* Stuessy
- 4. Series *Cupulata* Stuessy
  - 14, *M. cupulatum* A.Gray. 15, *M. appendiculatum* B.L.Rob. 16, *M. sinuatum* Brandegee. 17, *M. rosei* B.L.Rob. 18, *M. tenellum* Hook. & Arn. 19, *M. glabribracteatum* Stuessy
- 5. Series *Longipila* Stuessy
  - 20, *M. longipilum* B.L.Rob.

II. Section *Zarabellia* (Cass.) DC. ($x = 9$)
- 21, *M. longifolium* Cerv. ex Cav. 22, *M. mimulifolium* B.L.Rob. 23, *M. gracile* Less. 24, *M. microcephalum* Less. 25, *M. paniculatum* Gardner

III. Section *Serratura* Stuessy ($x = 12$)
- 26, *M. divaricatum* (Rich. in Pers.) DC. 27, *M. costaricense* Stuessy. 28, *M. dicoelocarpum* B.L.Rob. 29, *M. tepicense* B.L.Rob. 30, *M. sinaloense* Stuessy

IV. Section *Bibractiaria* Stuessy
- 31, *M. bibracteatum* S.Watson. 32, *M. repens* Sessé & Moç.

V. Section Rhizomaria Stuessy ($x = 11$)
- 33, *M. montanum* Benth. 33a, var. *montanum*. 33b, var. *viridulum* Stuessy. 34, *M. aureum* Brandegee

VI. Section *Alcina* (Cav.) DC. ($x = 11$)
- 35, *M. perfoliatum* (Cav.) Kunth. 36, *M. glabrum* S.Watson. 37, *M. nutans* Stuessy

(now known as $x = 14$), but their diminutive vegetative and floral morphology gave them a very distinctive appearance.

In the ensuing years, Turner added three additional species to the genus: *M. northingtonii* (1988) in Sect. *Serratura*, *M. mayfieldii* (1993) in Sect. *Melampodium* Ser. *Melampodium*, and *M. moctezumum* (2007) in Sect. *Melampodium* Ser. *Cupulata*. He also described another new species, *M. ellottianum* (2016), but this appears to be only a five-phyllary variant of the common *M. microcephalum* (see Stuessy, 1972: 168). With these new additions, there are now 40 species in the genus.

In the 1970s and 1980s, many phenetic studies were conducted on different genera of flowering plants as tests of the efficacy of the methods and to gain additional insights on relationships. Phenetics (numerical taxonomy) was pioneered by Robert Sokal and Peter Sneath (e.g., Sokal & Sneath, 1963; Sneath & Sokal, 1973) as a more objective means of constructing classifications. The approach advocated using as many characters as practicable, ideally more than 100, and deriving a quantitative measure of relationships based on a statistic. The idea was to divorce the process of classification from the personal biases of the individual investigator. Two principal methods had evolved: (1) cluster analysis and (2) ordination. The first sought to build trees (phenograms) by linking taxa based on levels of affinity, beginning with the highest values. The second sought maximum correlation among characters and plotted the samples in two-dimensional space based on the most prominent factors. These different approaches were often compared and contrasted for additional insights.

Phenetic methods were applied within *Melampodium* (Stuessy & Crisci, 1984) as a test of the hypotheses of the sectional and series classification by Stuessy (1972). At this time only 37 species had been described in the genus, and these served as the Operational Taxonomic Units (= OTUs) for the analyses. A total of 41 vegetative and reproductive characters were selected, in which character states could be non-arbitrarily delimited. Most features were qualitative, but some were quantitative (direct measurements). The cluster analysis that had the least distortion from the basic data matrix (i.e., the cophenetic correlation coefficient; Sokal & Rohlf, 1962) was based on both reproductive and vegetative characters (Fig. 10.6A). Both cluster analysis and ordination (Fig. 10.6B) revealed support for some of the infrageneric units, such as Sect. *Bibractiaria* (species 31 and 32), the white-rayed complex (species 6–8), and Sect. *Rhizomaria* (species 33 and 34). Most of the species of the large Sect. *Melampodium* clustered together, except for the inclusion of *M. dicoelocarpum*. Also clear were the weak sectional boundaries for Sects. *Zarabellia*, *Serratura* and *Alcina*. The latter was especially problematic, held together primarily by chromosome number ($n = 11$), but obviously with three morphologically diverse species. The bottom line from these phenetic studies was that distinct groupings morphologically do exist within the genus, but that also many parallelisms in morphology have taken place during the evolution of the group. Clearly *Melampodium* has undergone adaptive radiation, especially within Mexico, and this has also been accompanied by remarkable chromosomal change.

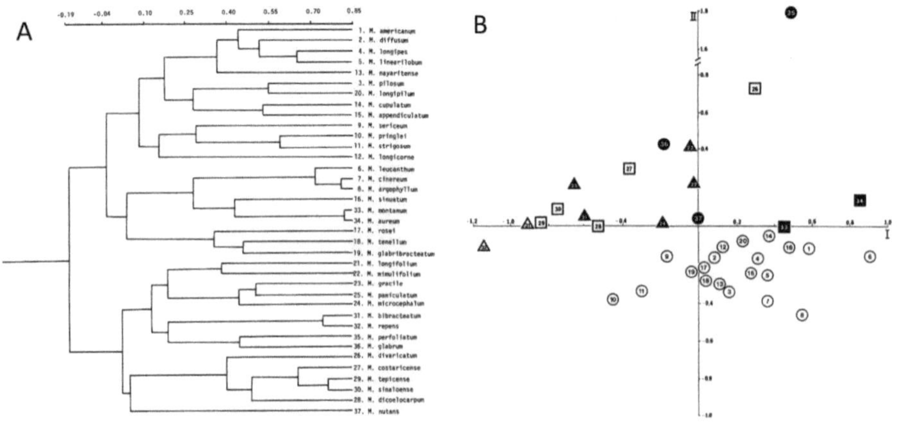

Fig. 10.6 Phenetic analyses of 37 species of *Melampodium* based on 42 reproductive and vegetative characters. (**A**) Cluster analysis; (**B**) principal components analysis. (From Stuessy & Crisci, 1984: 12)

A more significant test of the infrageneric classification proposed by Stuessy (1972) came with molecular phylogenetic analyses (Blöch et al., 2009; see Chap. 6). By this time, 40 species were known in the genus, and leaf material was obtained for DNA analysis from all of them, using nrITS, 5S rDNA spacer, low-copy nuclear *Pgi*C gene, and plastid regions *mat*K and *psb*A-*trn*H. Using species of *Milleria*, *Smallanthus*, and *Galinsoga* as outgroups, the results from ITS and *mat*K can be compared in Fig. 10.7. A number of similarities prevail, but also with some differences. The large Sect. *Melampodium* is a distinct group in both analyses, as are also Sect. *Serratura*, Sect. *Bibractiaria*, and Sect. *Rhizomaria*. Sect. *Zarabellia* is also relatively cohesive, except for *M. mimulifolium* and *M. longifolium*, which form a separate unit. The former is extremely similar morphologically to *M. gracile* of that same section, so much so that Stuessy (1972) opined that further collecting might recommend synonymizing the two, but the molecular data demonstrate distinct specific status.

The new insights from molecular phylogenetic analyses of *Melampodium* focus on the relationships of species in the former Sects. *Alcina*, *Serratura*, and *Zarabellia*. The molecular data reveal *M. nutans* to be divergent in both nuclear and chloroplast sequences (Fig. 10.7), as is also the case with *M. perfoliatum*, and the two are not at all closely related. Perhaps more surprising is that *M. glabrum*, which from the similarity of large ovate bracts and dark smooth fruits seemed close morphologically to *M. perfoliatum*, is quite divergent. The molecular results also reveal *M. nutans* to have no close relatives in any of the sections, and *M. perfoliatum* is likewise orphaned. *M. longifolium* joins with *M. mimulifolium*, but both fall somewhat outside the remainder of species of Sect. *Zarabellia*.

The two related genera, *Acanthospermum* and *Lecocarpus*, fall inside *Melampodium* as a group (Fig. 10.7), which might suggest submerging these two genera into *Melampodium* (as did Baillon, 1882), perhaps as a subgenus or section,

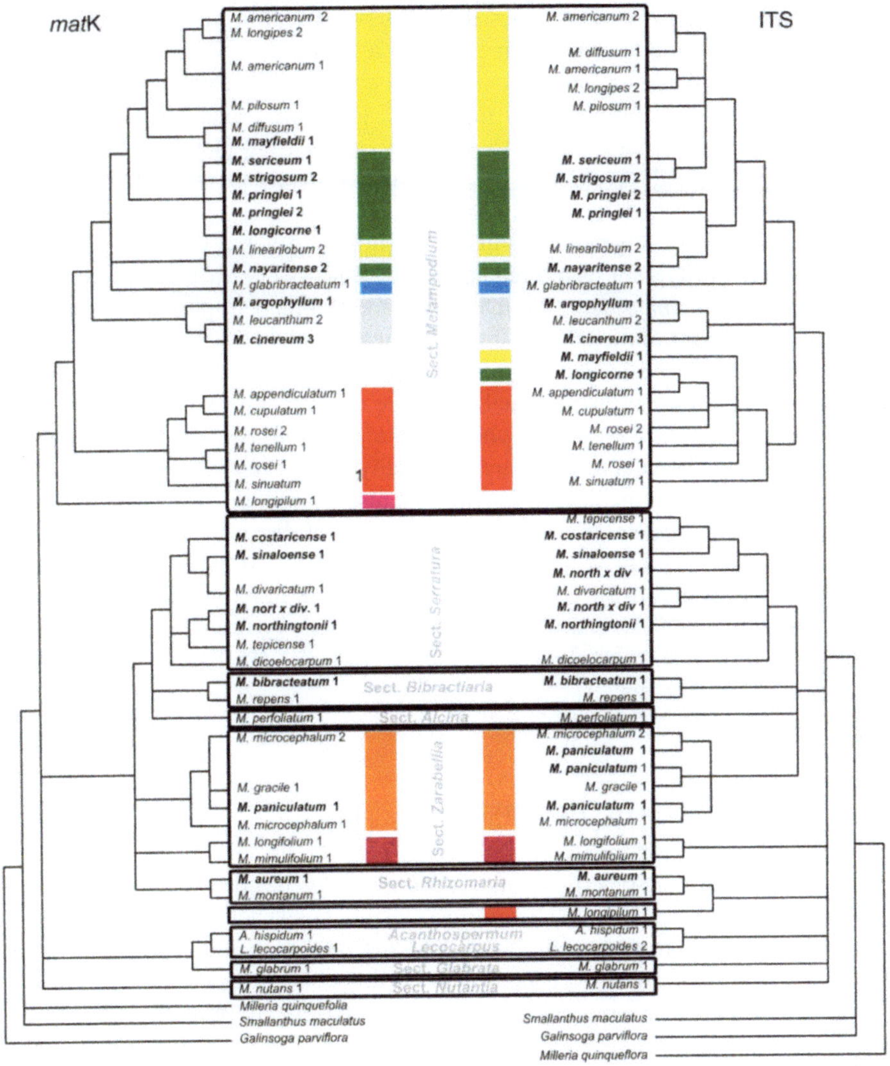

Fig. 10.7 Phylogenetic relationships among and within sections of *Melampodium* based on ITS and *mat*K data and analyzed with Bayesian inference displayed on 95% majority rule consensus trees. Colors show taxonomic series: pink, *Longipila*; purple, *Zarabellia*; brown, *Tribracteata*; red, *Cupulata*; green, *Sericea*; yellow, *Melampodium*; gray, *Leucantha*; blue, *Glabribracteata*. Numbers after species names refer to different accessions. Polyploid taxa are indicated in bold. (From Stuessy et al., 2011: 442, modified from Blöch et al., 2009)

but this would enlarge the concept of the genus for no particular gain except to preserve cladistic holophyly (for a full discussion of these issues, see Stuessy, 2009, 2025). In the studies of Rauscher (2002; Fig. 6.3), these two genera formed a sister group relationship with *Melampodium*. In an evolutionary context, *Melampodium* might have been the ancestral genus that diversified within Mexico, followed by further divergent evolution into a lineage that migrated into Central America and northern South America and eventually acquired characteristics resembling *Acanthospermum*. Long-distance dispersal from an ancestral species of *Acanthospermum* (perhaps similar to *A. microcarpum* from mainland Ecuador) to the Galápagos Islands eventually led to strongly divergent speciation into the shrubby *Lecocarpus* with its several island endemic species.

The molecular phylogenetic results have suggested some taxonomic changes (Table 10.2) to the hypothesis offered initially by Stuessy (1972; Table 10.1). Within the large Sect. *Melampodium*, the five series are maintained, plus the addition of Ser. *Glabribracteata* to include the single species, *M. glabribracteatum*, which had previously been placed in Ser. *Cupulata*. Sect. *Zarabellia* remains with the same specific composition, but two internal series are now recognized, Ser. *Tribracteata* to include *M. gracile*, *M. microcephalum*, and *M. paniculatum*, which from crossing studies (Stuessy & Brunken, 1979) are genetically closely related (see Chap. 8), and now Ser. *Zarabellia* to include *M. longifolium* and *M. mimulifolium*. The most significant changes in the classification are the creation of three new sections designed to deal with the divergent *M. nutans*, *M. perfoliatum*, and *M. glabrum*. The former is now the sole species in Sect. *Nutantia*, *M. perfoliatum* in its own Sect. *Alcina*, and *M. glabrum* alone in Sect. *Glabrata*. This new hypothesis of classification has eight sections (instead of six) and eight series (instead of five), which in the future can be tested with other datasets.

Specific and Infraspecific Taxa

Also significant among investigations of *Melampodium* has been the testing of limits of species. In general, more has been written about species than about any other level of the taxonomic hierarchy (e.g., Wilkins, 2009a, b, 2023; Zachos, 2016; Pavlinov, 2023), and renewed conceptual interest seems to cycle through the systematic and evolutionary biology community approximately once each generation. More than 20 different concepts of species have been proposed for different groups of organisms (Mayden, 1997). This book is not the place to reopen all the facets of opinion and perspective on this topic; for useful reviews, see Stuessy (2009) and Stuessy et al. (2014). Some comments, however, are believed appropriate.

Our general view of species in *Melampodium* adheres largely to the biological species concept (Dobzhansky, 1937; Mayr, 1957), which stresses reproductive isolation between interbreeding sets of populations. Even with modern molecular methods, in most cases we usually do not know the actual levels of gene flow or even actual genetic patterns within and among natural populations. Inferences must be made based on morphology, distribution, and ecology on what maintains the integrity of the observed taxa, which is similar to the morpho-genetic or ecogenetic concept of Levin (2000). In practice and with experience, determining species within *Melampodium* is not difficult. There are 40 species, and some of their

Table 10.2 Current classification of *Melampodium*

- I. Section *Rhizomaria* Stuessy ($x = 11$)
  - 1, *M. montanum* Benth. 1a, var. *montanum*. 1b, var. *viridulum* Stuessy. 2, *M. aureum* Brandegee (6*x*)
- II. Section *Glabrata* Stuessy ($x = 11$)
  - 3, *M. glabrum* S.Watson
- III. Section *Zarabellia* (Cass.) DC. ($x = 9$)
  - 1. Series *Zarabellia* Cass.
    - 4, *M. longifolium* Cerv. ex Cav. 5, M. *mimulifolium* B.L.Rob.
  - 2. Series *Tribracteata* Stuessy
    - 6, *M. microcephalum* Less. 7, *M. gracile* Less. 8, *M. paniculatum* Gardner (4*x*, 6*x*)
- IV. Section *Alcina* (Cav.) DC. ($x = 11$)
  - 9, *M. perfoliatum* (Cav.) Kunth
- V. Section *Bibractiaria* Stuessy ($x = 14$)
  - 10, *M. bibracteatum* S.Watson (4*x*). 11, *M. repens* Sessé & Moç. (2*x*, 4*x*)
- VI. Section *Serratura* Stuessy (x = 12)
  - 12, *M. divaricatum* (Rich. in Pers.) DC. 13, *M. costaricense* Stuessy (4*x*). 14, *M. northingtonii* B.L.Turner (4*x*). *M. tepicense* B.L.Rob. 16, *M. sinaloense* Stuessy (4*x*). 17, *M. dicoelocarpum* B.L.Rob. (2*x*, 4*x*-2)
- VII. Section *Nutantia* Stuessy ($x = 11$)
  - 18, *M. nutans* Stuessy
- VIII. Section *Melampodium* ($x = 10$)
  - 1. Series *Longipila* Stuessy
    - 19, *M. longipilum* B.L.Rob.
  - 2. Series *Cupulata* Stuessy
    - 20, *M. cupulatum* A.Gray. 21, *M. appendiculatum* B.L.Rob. 22, *M. moctezumum* B.L.Turner. 23, *M. sinuatum* Brandegee. 24, *M. rosei* B.L.Rob. 25, *M. tenellum* Hook. & Arn.
  - 3. Series *Leucantha* Stuessy
    - 26, *M. leucanthum* Torr. & A.Gray (2*x*, 4*x*). 27, *M. cinereum* DC. (2*x*, 4*x*). 27a, var. *cinereum*. 27b, var. *hirtellum* Stuessy. 27c, var. *ramosissimum* (DC.) A.Gray. 28, *M. argophyllum* (A.Gray ex B.L.Rob.) S.F.Blake (6*x*)
  - 4. Series *Glabribracteata* Stuessy
    - 29, *M. glabribracteatum* Stuessy
  - 5. Series *Sericea* Stuessy
    - 30, *M. sericeum* Lag. (6*x*). 31, *M. pringlei* B.L.Rob. (6*x*). 32, *M. strigosum* Stuessy (4*x*). 33, *M. nayaritense* Stuessy (4*x*). 34, *M. longicorne* A.Gray (6*x*)
  - 6. Series *Melampodium*
    - 35, *M. americanum* L. 36, *M. diffusum* Cass. 37, *M. pilosum* Stuessy. 38, *M. longipes* (A.Gray) B.L.Rob. 39, *M. linearilobum* DC. 40, *M. mayfieldii* B.L.Turner (4*x*)

From Stuessy et al. (2011)

Order of taxa follows primarily that from ITS analyses (Blöch et al., 2009: 225; see Figs. 6.4 and 6.7). Base chromosome numbers for sectional lineages and ploidy levels for polyploids are indicated

distinguishing features can be subtle, but with experience (especially in the field), they all are discernable.

Regarding infraspecific taxa, because only varieties have been employed historically within *Melampodium*, this is the only lower-level category of concern. Our view is that varieties must have morphological and geographical cohesion. That is, a variety must be mappable, such as seen among the three varieties of *M. cinereum* (Fig. 9.5). It is assumed that such units will be distinguishable genetically as well, as has been demonstrated in this same complex (Fig. 9.10). The only other varieties recognized formally in the genus are in *M. montanum* (vars. *montanum* and *viridulum*), but these have been much less collected and are not well understood geographically (Fig. 4.4B). It might be possible to formally resolve varieties within the extremely morphologically variable and widespread *M. americanum* (Fig. 4.3B), but such an effort has not yet been attempted.

Chapter 11
Conclusions

Research on *Melampodium* in Latin America and the southwestern United States over the past five decades has yielded useful insights for interpreting evolutionary relationships and mechanisms of speciation. In many ways, the genus is not so very different from other plant groups in these same geographic regions, but one particular aspect is noteworthy: *Melampodium* has undergone large amounts of chromosomal change during its evolutionary history with $n = 9$, 10, 11, 12, 14, 18, 20, 23, 24, 27, 28, 30, and 33.

Based on morphology, distributions, and chromosome numbers, a predictive classification of species into formal sections and series was proposed by Stuessy (1972). This served as a hypothesis of relationships that could be tested by new methodologies that developed in the 1970s and later. Three new species were added to the genus in subsequent decades (Turner, 1988, 1993, 2007). The more quantitative phenetic (Stuessy & Crisci, 1984) and cladistic (Stuessy, 1979) analyses in *Melampodium* largely confirmed the structure of sections except for Sect. *Alcina*. Although the three species in this section, *M. perfoliatum*, *M. glabrum*, and *M. nutans*, were all $n = 11$, the explicit methodologies indicated that this section was not a natural evolutionary unit (i.e., it was not monophyletic).

Further advances in understanding evolutionary relationships in *Melampodium* came with cytogenetic methodologies. A broadening of sampling of taxa, including expanding investigations of populations within the same species, led to a deeper understanding of variation within and among taxa (Stuessy et al., 2004; Weiss-Schneeweiss et al., 2009). A new base number was discovered in Sect. *Bibractiaria*: $x = 14$ in *M. repens* ($n = 14$ and 28) and *M. bibracteatum* ($n = 28$). Artificial hybridizations among the three species based on $x = 9$ (*M. gracile*, $n = 9$; *M. microcephalum*, $n = 9$; *M. paniculatum*, $n = 18$, 27) helped to clarify the allopolyploid origin of the tetraploid (Stuessy & Brunken, 1979).

Major improvements in understanding relationships in *Melampodium* came with the availability of DNA sequences and algorithms for phylogenetic analyses. Analyses of nuclear ITS and chloroplast *mat*K with all species led to new insights

T. F. Stuessy et al., *Systematics, Ecology, and Chromosomal Evolution of* Melampodium *(Asteraceae)*, https://doi.org/10.1007/978-3-031-81995-7_11

on phylogeny (Blöch et al., 2009). The basic assignment of species into taxonomic sections was confirmed, which also underlined the suitability of reliance on chromosome numbers and morphology for classification within the genus (Stuessy et al., 2011). Based on the new molecular phylogenetic results, a restructuring of the classification led to the breakup of Sect. *Alcina* into three distinct sections: *M. perfoliatum* remaining in Sect. *Alcina*, *M. glabrum* in a new Sect. *Glabrata*, and *M. nutans* in a new Sect. *Nutantia*. The species of Sect. *Zarabellia*, $x = 9$, were divided up into two distinct series. Five series in the largest Sect. *Melampodium* recognized by Stuessy (1972) remained the same, but *M. glabribracteatum* was segregated into its own series.

Armed with the new classification, it was possible to examine more carefully the chromosomal changes within the genus. Earlier chromosomal studies (Stuessy, 1971a) had suggested that $x = 10$ was most likely the ancestral base number for the genus. More recent molecular studies (McCann et al., 2016) have alternatively recommended that $x = 11$ is more probable as the generic base, in part because the related genera, *Acanthospermum* and *Lecocarpus*, are also based on this same number. These studies also indicated the occurrence of both descending dysploidy to $x = 10$ and $x = 9$ and also ascending dysploidy to $x = 12$ and $x = 14$.

From these new studies, it was possible to examine the origin of the polyploids that were derived from the base lineages. The origin of species by autopolyploidy has been hypothesized to have occurred in several species (e.g., *M. aureum*, $n = 33$), but this has been much less significant in the production of new species than by allopolyploidy (Weiss-Schneeweiss et al., 2012; McCann et al., 2018). In Sect. *Melampodium*, based on $x = 10$ that contains more than half of the species of the genus (22 of 40 total), nine different species are allotetraploid or allohexaploid. Most of these occur in Ser. *Sericea* (five species); the evolutionary connections among them are a fascinating and complex story (see Fig. 8.5).

Of particular interest has been the origin of repetitive DNA within the genome of species within the genus. Because repetitive DNA sequences are common in angiosperms (Pellicer et al., 2018), it was no surprise to learn that this is also the case in *Melampodium* (McCann et al., 2018). The role that these play in the genome of different species is not clear, but it was determined that variation in repetitive sequences does have phylogenetic import, as shown by different series in Sect. *Melampodium* having distinct repeat profiles.

The more precise phylogenetic inferences have allowed biogeographic hypotheses for *Melampodium* to be formulated. Regarding the entire genus, especially in Mexico where the concentration of species is the greatest, the center of present distribution is along the western central portion where three Cordilleras meet, which could be a center of geographic origin for the genus (suggested as the ancestral region; see Chap. 9). More species are found in pine-oak forests than in any other vegetation zone, and detailed analyses have revealed this to be the ancestral habitat (see Chap. 9). Phylogeographic studies have been done on the white-rayed complex in the southwestern United States and adjacent northern Mexico (Rebernig et al., 2010a, b, 2012). These population-level investigations were able to construct a new hypothesis, in conjunction with phylogenetic evidence with species in other series, that the white-rayed complex originated in northwestern Mexico and into Arizona, from an unknown ancestor in Ser. *Cupulata*, followed by eastern migration and differentiation into the three species seen today. This was in contrast to the earlier hypothesis (Stuessy, 1971b) of an origin on the northeastern side of Mexico followed by westward migration.

Every organismic group has evolved via some type of speciation. Characters have changed through time, proffering information for understanding evolutionary patterns and processes; the challenge is to investigate significant aspects within a monophyletic unit, whether genus, subgeneric unit, or species complex. In *Melampodium* this means investigating chromosome numbers and genomic structure to reveal relationships and unravel evolutionary mechanisms. This is not the only genus that has undergone substantial karyotypic modifications during its evolution, but it is one of the best studied examples among Asteraceae. There is still much to be learned from *Melampodium*, but the results in this book provide a new platform from which to continue to understand biodiversity within the genus.

Appendices

Appendix 1: Ploidy Level Determinations of Individuals in Populations of the White-Rayed Complex of *Melampodium*

Collected in silica gel, as determined by Renate Obermayer with flow cytometry following procedures in Doležel et al. (2007). Asterisks indicate populations in which ploidy level for some individuals (1–6) was reported already by Rebernig et al. (2010a and b*, 2012**). See footnote for names of collectors.

			Ploidy level					Coordinates	
Taxon	Population number	Number of plants	2*x*	3*x*	4*x*	5*x*	6*x*	Latitude	Longitude
M. argophyllum									
	*19059***	29					29	26° 14′ 55.2″	100° 41′ 22.4″
	*19060***	6					6	26° 13′ 54.9″	100° 40′ 38.1″
Subtotal	2	35					35		
M. cinereum									
var. ***cinereum***									
	18688	9	9					28° 46′ 5.2″	99° 16′ 12.08″
	*18689***	5	5					28° 36′ 8.2″	99° 50′ 7.55″
	*18690***	12	12					28° 02′ 1.9″	99° 31′ 11.8″
	*18691***	3	3					27° 46′ 14.1″	99° 27′ 1″
	18692	1	1					27° 56′ 8.3″	99° 28′ 1.05″
	*18693***	3	3					26° 53′ 6.95″	99° 15′ 11.6″

T. F. Stuessy et al., *Systematics, Ecology, and Chromosomal Evolution of* Melampodium *(Asteraceae)*, https://doi.org/10.1007/978-3-031-81995-7

(continued)

			Ploidy level					Coordinates	
Taxon	Population number	Number of plants	2*x*	3*x*	4*x*	5*x*	6*x*	Latitude	Longitude
	*18694***	21			20		1	26° 41′ 13.5″	99° 06′ 9.3″
	18696	3	3					26° 16′ 0.3″	98° 32′ 9.8″
	*18697***	31			31			26° 48′ 2.7″	98° 51′ 5.06″
	*18698***	22			21	1		26° 56′ 8.15″	98° 37′ 2.5″
	*18699***	11			11			27° 11′ 12″	98° 37′ 5.08″
	*18700***	10			10			27° 15′ 14.4″	98° 29′ 2.8″
	*18701***	26			25	1		27° 16′ 11″	98° 44′ 0.03″
	*18702***	6			6			27° 13′ 15.7″	98° 56′ 0.6″
	18703	12			12			27° 13′ 5.7″	98° 40′ 10.5″
	*18704***	15			15			27° 24′ 4.5″	98° 34′ 9.6″
	*18705***	14			14			27° 30′ 12.7″	98° 28′ 13.9″
	*18706***	14			14			27° 43′ 15.8″	98° 15′ 13.15″
	*18707***	26			26			27° 53′ 3.2″	98° 05′ 4.26″
	*18708***	26			24	2		28° 14′ 8.56″	98° 06′ 6.18″
Subtotal	20	270	36	0	229	4	1		
var. ***hirtellum***									
	*19054***	9	9					27° 56′ 44.8″	101° 15′ 24.3″
	*19055***	28	28					27° 54′ 10.4″	101° 20′ 33.8″
	*19057***	31	31					28° 11′ 36.0″	101° 04′ 18.7″
	*19058***	28	27	1				26° 30′ 31.4″	101° 21′ 06.4″
	*19061***	27	27					26° 21′ 56.7″	100° 07′ 22.6″
	*19062***	15	15					26° 37′ 34.2″	100° 01′ 32.6″
Subtotal	6	138	137	1					
var. ***ramosissimum***									
	18695	3	3					26° 26′ 5.3″	98° 43′ 2.3″
	*19063***	7	7					25° 10′ 48.4″	98° 11′ 56.1″
	19064	6	6					24° 51′ 46.9″	98° 08′ 21.0″

			Ploidy level					Coordinates	
Taxon	Population number	Number of plants	2*x*	3*x*	4*x*	5*x*	6*x*	Latitude	Longitude
	*19065***	24	24					24° 48′ 59.3″	98° 11′ 0.6″
Subtotal	4	40	40						
M. leucanthum (Texas)									
	18675	23			23			30° 13′ 14.7″	97° 55′ 13.5″
	*18676**, **	5	4		1			30° 11′ 10.41″	98° 14′ 5.68″
	*18677**, **	4	4					30° 12′ 58.68″	98° 14′ 18.2394″
	18678	4	4					30° 12′ 16.3″	98° 14′ 5″
	*18679**, **	12	12					30° 13′ 10.1″	98° 22′ 15.3″
	18680	2			2			30° 18′ 5.8″	98° 23′ 11.2″
	*18681**, **	9	9					30° 16′ 13.5″	98° 54′ 6.6″
	*18682**, **	21	20	1				30° 10′ 3.1″	98° 22′ 15.4″
	*18683**, **	15			15			29° 59′ 7.7″	98° 24′ 7.6″
	18684	25		1	24			29° 57′ 4.7″	98° 24′ 10.6″
	18685	14			14			29° 42′ 11.3″	98° 26′ 14.9″
	*18686**, **	26			26			29° 36′ 16.11″	98° 45′ 6.78″
	*18687**, **	14			14			29° 31′ 12.46″	98° 50′ 7.5″
	*18709**, **	13			13			30° 36′ 16.3″	97° 51′ 9.8″
	*18710**, **	25			25			30° 41′ 11.5″	97° 58′ 14.3″
	*18711**, **	15			15			30° 45′ 7.8″	98° 15′ 3.6″
	*18712**, **	25			25			30° 40′ 4.7″	98° 15′ 5.8″
	18713	10			10			30° 29′ 1.3″	98° 18′ 16.03″
	*18714**, **	10			10			30° 23′ 3.6″	98° 21′ 16.4″
	18715	23			23			30° 27′ 0.6″	98° 08′ 0.5″
	18716	15			15			30° 23′ 4.6″	97° 54′ 5.8″
	18717	26	26					29° 31′ 3.08″	99° 43′ 14.1″
	18718	7	7					30° 22′ 23.1″	101° 06′ 53.6″
	18719	15	15					29° 26′ 59″	100° 54′ 33.9″

(continued)

(continued)

			Ploidy level					Coordinates	
Taxon	Population number	Number of plants	2*x*	3*x*	4*x*	5*x*	6*x*	Latitude	Longitude
	*18720**, **	5			5			29° 43′ 6.2″	101° 21′ 39.6″
	18721	1	1					30° 10′ 2.8″	102° 40′ 53.5″
	*18722**	9	9					30° 44′ 53.5″	102° 54′ 23.9″
	18723	6	6					30° 40′ 39.7″	102° 56′ 7.4″
	18724	2	2					30° 12′ 24.1″	103° 14′ 15.1″
	*18725**, **	25	25					30° 16′ 46.5″	104° 01′ 59.8″
	*18726**, **	8	7	1				31° 00′ 19.5″	104° 49′ 30.9″
	*18727**, **	13	13					30° 59′ 56.8″	103° 45′ 20.7″
	*18729**, **	16	16					31° 33′ 8.5″	103° 34′ 47.5″
	*18730**, **	15	15					31° 38′ 23.1″	102° 59′ 2.6″
	*18731**, **	8	7	1				31° 41′ 52.7″	102° 34′ 27.6″
	*18732**, **	1	1					31° 56′ 4.1″	101° 51′ 55.9″
	*18733**, **	9	9					31° 52′ 15.7″	101° 38′ 45.7″
	*18734**, **	6	6					31° 53′ 59.1″	100° 43′ 0.2″
	*18735**, **	16	16					32° 57′ 50.3″	102° 00′ 42.4″
	18736	26	26					32° 26′ 39.3″	102° 19′ 49.4″
	*18737**, **	11	11					32° 17′ 15.3″	102° 36′ 40.8″
	*18738**, **	15	15					31° 51′ 6.3″	103° 06′ 51.0″
	18739	8	8					31° 46′ 15.3″	103° 25′ 20.4″
	18740	4	4					31° 41′ 52.0″	103° 36′ 19.2″
	18741	5	5					32° 59′ 50.6″	104° 26′ 24.9″
	18742	16	15	1				33° 25′ 11.0″	104° 00′ 0.5″

			Ploidy level					Coordinates	
Taxon	Population number	Number of plants	2*x*	3*x*	4*x*	5*x*	6*x*	Latitude	Longitude
	18743	25	25					33° 10′ 55.1″	102° 43′ 33.4″
	18744	13	13					33° 53′ 17.9″	102° 33′ 27.6″
	18745	1	1					34° 49′ 19.3″	103° 22′ 58.7″
	18746	7	7					34° 49′ 50.6″	103° 45′ 15.0″
	18747	6	6					35° 20′ 41.6″	103° 23′ 29.4″
	18748	15	15					36° 20′ 40.6″	104° 15′ 31.8″
	18749	7	7					36° 19′ 12.9″	104° 28′ 39.5″
	18750	26	26					37° 57′ 34.3″	104° 46′ 43.5″
	18751	9	9					36° 02′ 21.3″	102° 30′ 17.6″
	18752	15	15					36° 18′ 31.0″	102° 03′ 44.8″
	*18753**, **	4	4					35° 19′ 35.0″	102° 22′ 11.3″
	18754	5	5					35° 24′ 38.3″	102° 17′ 13.5″
	18755	2	2					35° 23′ 23.4″	102° 05′ 47.4″
	*18756**, **	25	25					35° 00′ 9.0″	101° 55′ 17.4″
	*18757**, **	3	3					34° 59′ 09.5″	101° 43′ 02.2″
	*18758**, **	10	10					36° 13′ 15.0″	101° 20′ 01.6″
	18759	5	5					36° 15′ 0.3″	100° 49′ 03.8″
	*18760**, **	4	4					36° 26′ 56.3″	100° 22′ 20.4″
	*18761**, **	14	12	2				36° 25′ 35.6″	99° 53′ 0.0″
	*18762**, **	4	4					35° 50′ 42.8″	100° 23′ 49.0″
	18763	6	6					35° 32′ 52.8″	101° 09′ 07.3″
	*18764**, **	4	4					35° 25′ 54.5″	100° 46′ 13.3″

(continued)

(continued)

Taxon	Population number	Number of plants	Ploidy level					Coordinates	
			2*x*	3*x*	4*x*	5*x*	6*x*	Latitude	Longitude
	18765	8	8					34° 43′ 05.5″	100° 16′ 22.8″
	18766	14	14					33° 37′ 35.2″	100° 50′ 33.9″
	18767	26	26					33° 48′ 08.5″	99° 47′ 36.7″
	*18768***	14	14					33° 51′ 34.9″	100° 51′ 05.8″
	*18769**, **	16	16					34° 13′ 11.4″	100° 53′ 16.9″
	*18770**, **	20	20					34° 22′ 46.6″	101° 06′ 39.0″
	*18771**, **	1	1					34° 47′ 18.1″	100° 53′ 54.0″
	*18772**, **	9	8	1				35° 00′ 31.2″	100° 53′ 44.8″
	18773	1			1			32° 04′ 13.0″	99° 41′ 14.3″
	*18774**, **	27	26	1				31° 14′ 08.1″	99° 45′ 25.2″
	18775	9	9					31° 10′ 43.4″	99° 24′ 27″
	*18776**, **	15			15			31° 46′ 52″	99° 21′ 45″
	18777	4			4			31° 53′ 53.1″	98° 54′ 53.4″
	*18778**, **	21			21			31° 45′ 41.2″	98° 53′ 56.5″
	*18779**, **	21			21			31° 41′ 18.9″	98° 48′ 48.5″
	18780	15			14		1	31° 26′ 39.6″	98° 33′ 59.8″
	*18781**, **	15			15			31° 21′ 27.8″	98° 18′ 46.7″
	*18782**, **	16			16			31° 01′ 11.2″	98° 11′ 11.9″
	*18783**, **	1			1			30° 55′ 42.3″	98° 00′ 05.9″
	18784	5			5			31° 04′ 40.7″	98° 01′ 42.7″
	*18785**, **	25			25			31° 03′ 51.9″	97° 34′ 18.5″
	*18786**, **	5			5			31° 34′ 31.5″	97° 50′ 03.0″

Taxon	Population number	Number of plants	Ploidy level					Coordinates	
			2*x*	3*x*	4*x*	5*x*	6*x*	Latitude	Longitude
	*18787**, **	3			3			31° 47′ 35.8″	98° 13′ 39.8″
Subtotals	91	1084	668	9	406	0	1		
M. leucanthum **(Arizona)**									
	*18800**, **	28	28					33° 14′ 18.816″	110° 15′ 11.52″
	*18801**, **	26	26					34° 0.0′ 8.568″	111° 18′ 49.212″
	*18802**, **	5	5					33° 28′ 49.188″	111° 26′ 35.52″
	18803	19	19					33° 53′ 54.204″	111° 49′ 21.252″
	18804	26	26					33° 55′ 34.788″	111° 48′ 56.952″
	*18805**, **	10	10					33° 57′ 46.008″	111° 51′ 46.62″
	18806	8	8					34° 14′ 13.488″	112° 6′ 34.056″
	*18807**, **	10	10					34° 20′ 56.904″	112° 9′ 13.824″
	*18808**, **	25	25					34° 36′ 20.484″	111° 51′ 27.94″
	*18809**, **	10	10					34° 37′ 3.0″	111° 50′ 35.48″
	*18810**, **	10	10					34° 39′ 0.108″	111° 45′ 36.54″
	*18811**, **	10	10					34° 38′ 21.3″	111° 48′ 27.40″
	*18812**, **	8	8					34° 42′ 43.164″	111° 52′ 50.52″
	*18813**, **	8	8					34° 43′ 47.928″	111° 58′ 7.968″
	*18814**, **	15	15					34° 45′ 18.72″	112° 5.0′ 29.04″
	18815	26	26					34° 44′ 50.316″	112° 6′ 20.844″
	*18816**, **	10	10					34° 49′ 31.908″	111° 46′ 42.85″
	*18817**, **	6	6					34° 45′ 24.37″	111° 45′ 54.072″
	18818	3	3					34° 45′ 24.408″	111° 45′ 54.04″
	*18819**, **	18	18					35° 31′ 18.19″	113° 27′ 19.044″

(continued)

(continued)

			Ploidy level					Coordinates	
Taxon	Population number	Number of plants	2*x*	3*x*	4*x*	5*x*	6*x*	Latitude	Longitude
	*18820**, **	16	16					35° 7.0′ 17.4″	113° 39′ 58.68″
	18821	3	3					34° 28′ 49.62″	113° 20′ 30.48″
	*18822**, **	17	17					32° 36′ 7.416″	110° 44′ 41.172″
Subtotals	23	317	317	0	0	0	0		
M. leucanthum (Arizona)									
	*20000**, **	30	30					32° 15′ 46.8’	107° 13′ 55.2″
	*20001**	16	16					32° 18′ 31.2984″	107° 17′ 34.577″
	*20002**, **	30	30					31° 57′ 10.8″	107° 40′ 30″
	*20003**, **	15	15					32° 38′ 24″	107° 57′ 14.4″
	20004	6	6					32° 40′ 43.14″	108° 7′ 49.573″
	*20005**, **	23	23					32° 47′ 9.6″	108° 8′ 20.4″
	20006	31	31					32° 55′ 7.284″	107° 41′ 51.18″
	*20007**, **	15	15					32° 57′ 10.8″	107° 29′ 20.4″
	20008	1	1					33° 14′ 45.96″	107° 16′ 5.484″
	20009	28	28					33° 17′ 6.755″	107° 27′ 10.367″
	*20010**, **	14	14					33° 16′ 29.99″	107° 16′ 51.6″
	*20011**, **	18	18					34° 8.0′ 45.6″	106° 54′ 25.199″
	20012	1	1					33° 54′ 39.697″	106° 52′ 46.559″
	20013	30	30					33° 52′ 52.057″	106° 26′ 29.076″
	*20014**, **	13	13					33° 47′ 34.799″	106° 16′ 26.4″
	20015	19	19					33° 43′ 28.739″	105° 50′ 14.352″
	*20016**, **	29	29					33° 0.0′ 32.4″	105° 56′ 27.6″
	*20017**, **	21	21					34° 56′ 42″	106° 11′ 27.6″

			Ploidy level					Coordinates	
Taxon	Population number	Number of plants	2*x*	3*x*	4*x*	5*x*	6*x*	Latitude	Longitude
	20018	21	20	1				34° 25′ 53.652″	106° 33′ 23.687″
	20019	2	2					34° 47′ 29.976″	106° 54′ 49.298″
	20020	9	9					35° 12′ 34.595″	106° 18′ 30.996″
	*20021***	28	28					35° 17′ 16.799″	106° 12′ 57.599″
	20022	15	15					35° 26′ 9.275″	105° 36′ 25.272″
	20023	4	4					35° 27′ 49.283″	105° 12′ 59.399″
	20024	30	30					35° 13′ 32.988″	105° 5′ 33.251″
	20025	26	26					36° 28′ 11.999″	104° 3′ 19.187″
	20027	29	29					35° 43′ 52.5″	103° 7′ 32.858″
	20028	19	19					35° 9′ 56.199″	103° 19′ 53.147″
	*20029**, **	24	24					35° 13′ 47.999″	103° 46′ 1.199″
	*20030**, **	27	27					35° 23′ 42″	104° 10′ 44.4″
	*20031**, **	26	26					34° 53′ 49.199″	104° 43′ 4.8″
	*20032***	13	13					34° 2′ 9.6″	104° 44′ 49.199″
	*20033**, **	6	6					32° 29′ 24″	104° 20′ 52.799″
	*20034**, **	27	27					32° 31′ 44.4″	103° 48′ 3.6″
	*20035**, **	30	30					32° 30′ 21.6″	103° 7′ 33.6″
Subtotals	35	676	675	1					
M. leucanthum (west Texas)									
	20037	9	9					31° 38′ 15.72″	102° 58′ 45.371″
	*20038**, **	13	13					30° 14′ 20.4″	103° 22′ 47.999″
	*20039**, **	15	14	1				29° 47′ 2.399″	103° 10′ 37.2″
	*20040**, **	3	3					29° 30′ 57.599″	103° 24′ 10.8″

(continued)

(continued)

			Ploidy level					Coordinates	
Taxon	Population number	Number of plants	2*x*	3*x*	4*x*	5*x*	6*x*	Latitude	Longitude
	20041	23	23					29° 47′ 1.248″	103° 34′ 38.028″
	20042	15	15					30° 16′ 50.879″	103° 54′ 6.479″
	*20043**, **	10	10					31° 0.0′ 10.8″	104° 49′ 26.399″
	*20044**, **	15	15					31° 36′ 35.99″	104° 51′ 25.199″
	*20045**, **	30	30					31° 18′ 54″	106° 4′ 40.8″
Subtotals	9	133	132	1					
M. leucanthum **(New Mexico)**									
	*20046**, **	29	29					32° 23′ 49.199’	106° 36′ 50.4″
M. leucanthum **(Coahuila, Mexico)**									
	*19056***	1	1					28° 13′ 44.5″	101° 02′ 38.1″
Grand totals	192	2723	2035	12	635	4	37		

Note: *18675–18688*, *18695*, *18708*, TS & CR; *18709–18787*, CR & ML; *18800–18822*, CR & MB; *19054–19062*, TS, JV, CR & CB; *19063–19065*, TS, JV & CB; *20000–20046*, CR & CB. *CB* C. Blöch, *CR* C. Rebernig, *JV* J. Villaseñor, *MB* M. Barfuss, *ML* M. Lenko, *TS* T. Stuessy

Appendix 2: Analysis of Ranges of Elevation for Species of *Melampodium* (Used in Chap. 8)

A total of 6359 georeferenced records from the *Melampodium* database were used to extract the elevation values. The Worldclim altitude layer was used (https://www.worldclim.org/data/worldclim21.html) at a resolution of 30 arc seconds. Spatial operations were performed in Arcmap 10.5 (ESRI, 2013). With the elevation values obtained, descriptive statistics were calculated: minimum, maximum, mean, and standard deviation (Table A.1; for boxplot graphics see Fig. 8.1). The statistical analyses were done in R 4.3.1 (R Core Team, 2024).

Descriptive statistics on elevation in *Melampodium* species. Values are in meters above sea level

	Minimum	Mean	Maximum	Standard deviation
Melampodium americanum	4	788.5	2175	577.2
Melampodium appendiculatum	173	1067.2	2405	473.4
Melampodium argophyllum	566	1299.9	2116	442.6
Melampodium aureum	846	2237.9	3008	410.9
Melampodium bibracteatum	1709	2602.8	3471	265.1
Melampodium cinereum	27	250.4	1586	283.3
Melampodium costaricense	10	537.2	1931	544.2
Melampodium cupulatum	5	267.5	760	270.7
Melampodium dicoelocarpum	158	1258.1	2042	518.7
Melampodium diffusum	3	158.6	1513	276.7
Melampodium divaricatum	−1	732.2	3235	672.0
Melampodium glabribracteatum	1501	1508.5	1517	6.6
Melampodium glabrum	1534	1914.0	2775	284.6
Melampodium gracile	−1	755.9	2062	610.0
Melampodium leucanthum	162	1184.5	3820	482.0
Melampodium linearilobum	6	560.9	2045	487.7
Melampodium longicorne	1045	1428.7	2202	230.7
Melampodium longifolium	194	2281.3	2746	400.5
Melampodium longipes	215	980.1	1632	390.8
Melampodium longipilum	22	1243.3	2466	570.8
Melampodium mayfieldii	957	1104.2	1168	74.1
Melampodium microcephalum	18	1347.6	2575	580.4
Melampodium mimulifolium	1634	2022.2	2332	261.7
Melampodium moctezumum	635	790.9	1400	269.6
Melampodium montanum	1	2138.3	3288	506.3
Melampodium nayaritense	339	738.0	1197	260.4
Melampodium northingtonii	58	329.8	561	206.7
Melampodium nutans	22	448.6	1726	448.9
Melampodium paniculatum	11	1075.9	2257	537.9
Melampodium perfoliatum	2	1647.5	2831	561.9
Melampodium pilosum	93	582.7	2329	475.8
Melampodium pringlei	1670	2011.7	2266	185.1
Melampodium repens	2053	2758.6	3416	348.0
Melampodium rosei	1	133.2	931	210.2
Melampodium sericeum	484	1768.4	2604	387.6
Melampodium sinaloense	688	1006.6	1661	377.0
Melampodium sinuatum	24	43.0	70	24.2
Melampodium strigosum	1067	1822.7	2513	326.1
Melampodium tenellum	4	148.3	942	248.2
Melampodium tepicense	62	1131.3	2175	446.4

Appendix 3: Methods for Estimation of Overlap of Ranges of Species of *Melampodium* (Used in Chap. 8)

The minimum convex hull has been widely used to estimate the distributional ranges of species (Meyer et al., 2017; Yalcin and Leroux, 2017). It is a spatial analysis technique that converts the points of the collection coordinates of the herbarium specimens into data delimited by area (a polygonal surface). The method uses the Euclidean distance between occurrences of the species to create a polygon that covers all occurrences without any internal angle exceeding 180°, minimizing the distribution surface of the points (Burgman & Fox, 2003).

For species of *Melampodium*, the occurrence records were taken to generate the minimum convex polygons for each of them and calculate their surface area. Subsequently, union operations were performed to identify the overlap between each pair of convex polygons, and their surface area in km^2 was calculated. All these operations were performed in the Geographical Information System ArcMap 10.5 (ESRI, 2013). Once the surface of the convex polygons and the overlap between each pair of species were obtained, this value was converted to a percentage taking as a reference the surface of the smallest polygon of each pair.

Appendix 4: Guide to Identification of Species of *Melampodium*

Artificial Key to Taxa

The following key to taxa of *Melampodium* is that of Stuessy (1972) plus additions of the newer species, *M. mayfieldii*, *M. moctezumum*, and *M. northingtonii*, more recently described by Turner (1988, 1993, 2007, respectively). The key is by intent artificial, i.e., the species are ordered in a manner most helpful for identification without attempting to reflect evolutionary affinities. Numbers by species names refer to their order in the new classification presented by Stuessy et al. (2011: 438), which follows the topological structure of the phylogenetic tree based on nr ITS sequences (from Stuessy et al., 2011: 442; Fig. 10.7; refer also to Table 10.2). In the key, "ligules" refer to corollas of the ray florets.

a. Ligules white.

  b. Outer phyllaries connate more than half their length.

    c. Stems and leaves strigillose; leaves with margins usually entire-sinuate (less often pinnately lobed); heads 20–37 mm diam;[1] ligules 7–13 mm long, 2.5–8 mm wide; plants 15–50 cm tall. . . . . 26. *M. leucanthum*

[1] Head diameter excludes the outer involucral bracts.

c. Stems and leaves canescent-tomentose; leaves with margins usually pinnately lobed-parted (less often entire); heads 10–23 mm diam; ligules 2.5–9 mm long, 1.5–4.5 mm wide; plants 12–22 cm tall. . . . . 28. *M. argophyllum*

b. Outer phyllaries connate a third or less their length.

d. Leaves with basal margins hispid with hairs 0.6–1.5 mm long; stems strigose-hispid with hairs 0.2–1.5 mm long. . . . . 27b. *M. cinereum* var. *hirtellum*

d. Leaves with basal margins strigose with hairs 0.3–0.6 mm long; stems strigose with hairs 0.1–0.8 mm long.

e. Leaves linear, 7–32 mm long, 2–5 mm wide; heads 3–5 mm tall; outer involucre 4.5-9 mm diam; outer phyllaries 2.7–4 mm long, 1.3–2.8 mm wide; hood 0.8–1.5 mm tall. . . . . 27c. *M. cinereum* var. *ramosissimum*

e. Leaves linear-oblong, 12–55 mm long, 1–14 mm wide; heads 5–8 mm tall; outer involucre 7–14.3 mm diam; outer phyllaries 3.5–7.3 mm long, 2.3–5 mm wide; hood 1.3–2.9 mm tall. . . . 7a. *M. cinereum* var. *cinereum*

a. Ligules yellow or yellow-orange.

f. Outer phyllaries 2.

g. Leaves oblong to obovate-oblanceolate; ray florets 5–6 (rarely 3); stems erect to decumbent; plants subaquatic. . . . . 10. *M. bibracteatum*

g. Leaves obovate; ray florets 2–3; stems prostrate; plants terrestrial. . . . . 11. *M. repens*

f. Outer phyllaries 3–5.

h. Outer phyllaries 3.

i. Heads 3–5.5 mm diam; ligules usually less than 2 mm long; peduncles copiously stipitate-glandular. . . . . 8. *M. paniculatum*

i. Heads 5–15 mm diam; ligules usually more than 2 mm long; peduncles strigillose to hispid to weakly stipitate-glandular.

j. Leaves subauriculate at base. . . . . 7. *M. gracile*

j. Leaves attenuate to obtuse at base.

k. Leaves with tapering petioles 1–3 cm long; inner phyllaries open in a small hood at achene apex, often extended into awn. . . . . 18. *M. nutans*

k. Leaves sessile or with petioles 0.1–0.4 cm long; inner phyllaries completely closed over achene apex, with no extended appendage. . . . . 6. *M. microcephalum*

h. Outer phyllaries 5.

l. Leaves sericeous beneath.

m. Outer phyllaries with the margin scarious. . . . . 39. *M. linearilobum*

m. Outer phyllaries with the margin herbaceous.

n. Leaves narrowly ovate to lanceolate.

o. Plants perennial; heads 11–17 mm diam; ligules 4–6 mm long. . . . . .35. *M. americanum*

o. Plants annual; heads 6–11 mm diam; ligules 2.5–4.5 mm long.

p. Plants 8–28 cm tall; stems pilose; outer phyllaries ovate to narrowly ovate; leaves copiously sericeous beneath. . . . . 37. *M. pilosum*

p. Plants 25–40 cm tall; stems strigillose-hispidulous; outer phyllaries ovate; leaves moderately sericeous beneath. . . . . 36. *M. diffusum*

n. Leaves linear, oblanceolate, or deeply pinnately divided.

q. Ligules 2 mm long or less.

r. Leaves linear and entire to deeply pinnately divided; ligules ca. 2 mm long; heads 8–11 mm diam; disc florets 15–25. . . . . 33. *M. nayaritense*

r. Leaves linear-oblong to oblanceolate, often lobed or parted; ligules ca. 1 mm long; heads 4–8 mm diam; disc florets 2–12.

s. Peduncles 5–22 mm long; disc florets 5–12; ligules on undersurface yellow at apex; outer involucre 6–11 mm diam. . . . . 30. *M. sericeum*

s. Peduncles 1–2 mm long; disc florets 2–3; ligules on undersurface purple at apex; outer involucre 3–4 mm diam. . . . . 31. *M. pringlei*

q. Ligules more than 2 mm long.

t. Plants annual; stems yellow to green. . . . . 38. *M. longipes*

t. Plants perennial; stems purple to green. . . . . 35. *M. americanum*

l. Leaves glabrous, strigose or pilose beneath.

u. Outer phyllaries with the margin narrowly scarious.

v. Peduncles 1.3–1.7 cm long; outer phyllaries abaxially glabrous. . . . . .29. *M. glabribracteatum*

v. Peduncles usually more than 2 cm long; outer phyllaries abaxially subglabrous, strigose or pilose.

w. Plants perennial with fibrous roots and rhizomes.

x. Heads 19–38 mm diam; ligules yellow-orange to yellow, 5–13 mm long; disc corollas yellow; fruit 2–2.3 mm long;[2] chromosome number, $n = 33$. . . . . 2. *M. aureum*

x. Heads 11–21 mm diam; ligules yellow, 4–7.5 mm long; disc corollas yellow-green; fruit 1.6–1.8 mm long; chromosome number, $n = 11$.

y. Ligules on undersurface light green at apex and on veins. . . . . 1b. *M. montanum* var. *viridulum*

y. Ligules on undersurface dark purple at apex and on veins. . . . . 1a. *M. montanum* var. *montanum*

w. Plants annual or perennial with tap roots.

z. Plants perennial; leaves on undersurface copiously tomentose or white-pilose.

aa. Leaves at margin sinuate-undulate, undersurface copiously strigillose or tomentose; margin of outer phyllaries slightly scarious; Baja California. . . . . 23. *M. sinuatum*

aa. Leaves at margin subentire, undersurface densely pilose; margin of outer phyllaries herbaceous (sometimes purple-scarious toward apex); Jalisco. . . . . 40. *M. mayfieldii*

z. Plants annual; leaves on undersurface strigose, lobed or divided.

bb. Leaves at base attenuate.

cc. Outer phyllaries ovate to narrowly ovate, 4–6 mm long; heads 7–9 mm tall.

dd. Leaves with margins lobed or irregularly serrate; outer phyllaries broadly ovate, connate ⅔–⅘ their length. . . . . . . .22. *M. moctezumum*

dd. Leaves with margins usually entire, sometimes crenate or sinuate; outer phyllaries ovate to narrowly ovate, connate ½ their length. . . . . 20. *M. cupulatum*

cc. Outer phyllaries ovate to orbiculate, 2.5–3.5 mm long; heads 3.5–5.5 mm tall. . . . . 25. *M. tenellum*

bb. Leaves at base obtuse to subauriculate.

ee. Outer phyllaries connate ⅔ their length. . . . . 21. *M. appendiculatum*

ee. Outer phyllaries connate less than ¼ their length. . . . . .24. *M. rosei*

[2] Fruit length excludes hoods and appendages.

u. Outer phyllaries with the margin herbaceous.

ff. Outer involucre 15–32 mm diam; outer phyllaries 6–20 mm long; fruits 4–7 mm long; lower leaves conspicuously perfoliate at base. . . . . 9. *M. perfoliatum*

ff. Outer involucre 3–10 (−14) mm diam; outer phyllaries 2–6 (−6.8) mm long; fruits 1.6–3 (−4) mm long; lower leaves attenuate to subauriculate at base.

gg. Leaves markedly petiolate.

hh. Ligules 3.5–7 mm long.

ii. Fruit at apex with a flattened adaxial appendage extending upward into a cirrhous awn (Fig. 3.3C). . . . . 19. *M. longipilum*

ii. Fruit at apex variously ridged and sculptured but not as above. . . . . .12. *M. divaricatum*

hh. Ligules less than 3 mm long.

jj. Fruit on lateral surfaces with 2 deep oval cavities. . . . . 17. *M. dicoelocarpum*

jj. Fruit on lateral surfaces variously ribbed, tubercled, and sculptured but not as above.

kk. Heads 6–10 mm diam; stems usually 2–4 mm diam. . . . . 13. *M. costaricense*

kk. Heads 3–5 mm diam; stems usually less than 2 mm diam.

ll. Peduncles 0.8–8 mm long; disc florets 3–7 (−12). . . . . 15. *M. tepicense.*

ll. Peduncles 8–29 mm long; disc florets 10–14.

mm. Leaf blades 3.5–5 cm long; heads 3–5 mm tall; involucre 5-10 mm diam; fruits at apex smooth or with tubercules; Sinaloa. . . . . .16. *M. sinaloense*

mm. Leaf blades 2–3.5 cm long; heads 2–3 mm tall; involucre 3.5-5 mm diam; fruits at apex tuberculate or sometimes with an abaxial awn to 2.5 mm long; Oaxaca. . . . . .14. *M. northingtonii*

gg. Leaves sessile or subsessile.

nn. Plants subaquatic; stems thick and yellowish. . . . . 3. *M. glabrum* nn. Plants terrestrial; stems greenish to greenish-purple.

oo. Ligules less than 2 mm long; outer phyllaries lanceolate.

pp. Fruits with an abaxial protuberance (Fig. 3.3B). . . . . 4. *M. longifolium*

pp. Fruits with hoods (Fig. 3.3D) or minutely sculptured at apex.

qq. Plants usually 15–20 cm tall; stems ascending; peduncles usually 0–3 mm long; leaves at margin entire-pinnately lobed; heads 3–4 mm tall, 4–8 mm diam; outer phyllaries lanceolate-elliptic. . . . . 32. *M. strigosum*

qq. Plants usually 20–40 cm tall; stems erect; peduncles usually 10–20 mm long; leaves at margin entire to dentate; heads 4–7 mm tall, 7–13 mm diam; outer phyllaries elliptic to obovate-rhombic. . . . . 34. *M. longicorne*

oo. Ligules more than 2 mm long; outer phyllaries ovate or narrowly ovate.

rr. Leaves attenuate at base. . . . . 6. *M. microcephalum*

rr. Leaves subauriculate to obtuse at base.

ss. Stems copiously pilose near apex; outer phyllaries ovate-lanceolate, 5–6 mm long, at apex acute. . . . . 5. *M. mimulifolium*

ss. Stems moderately hispid-pilose and weakly stipitate-glandular near apex; outer phyllaries ovate, 3.5–5 mm long, at apex acuminate. . . . . 7. *M. gracile*

Habit Photographs and Distribution Maps of All Species (Ordered Alphabetically)

For authorities of all taxa, see Table 10.2

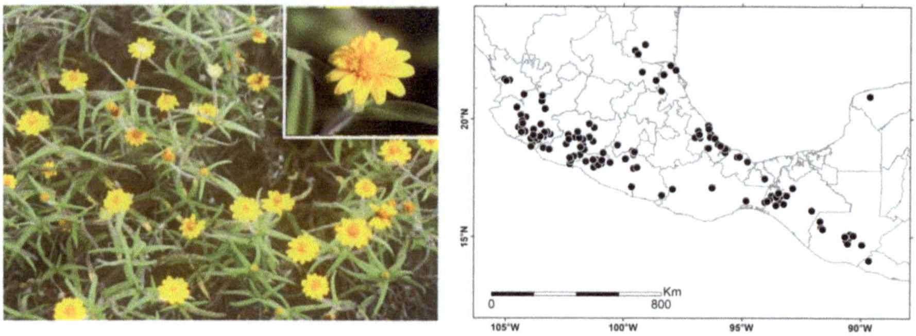

M. americanum

M. appendiculatum

M. argophyllum

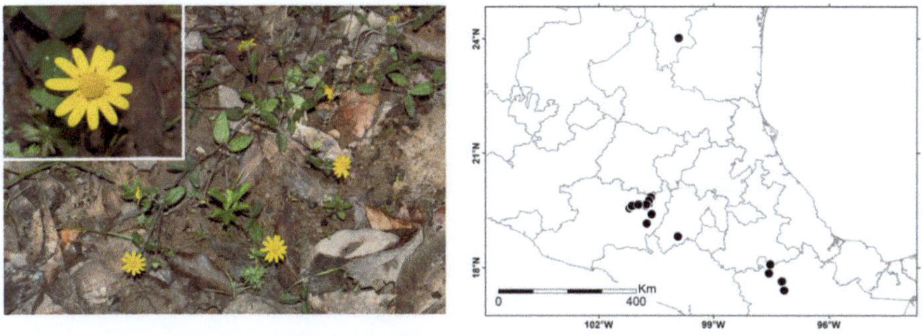

M. aureum

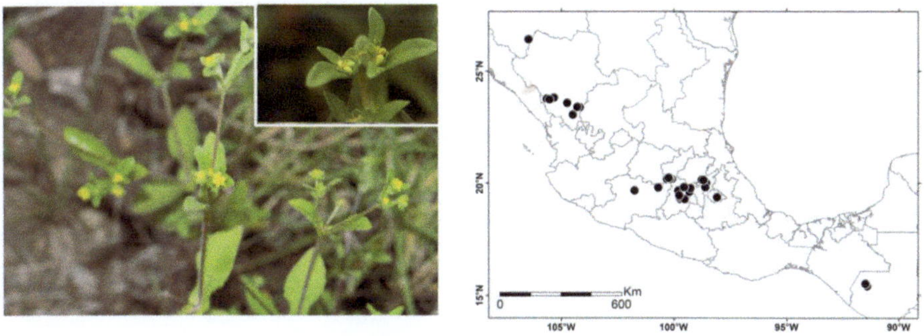

M. bibracteatum

M. cinereum (var. *cinereum,* circles; var. *hirtellum,* squares; var. *ramosissimum,* triangles)

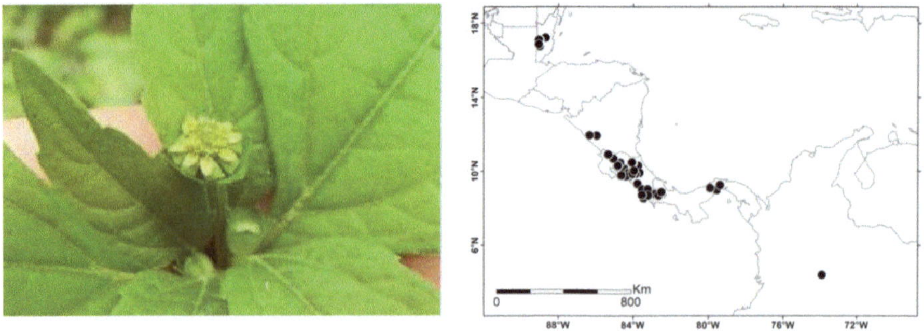

M. costaricense

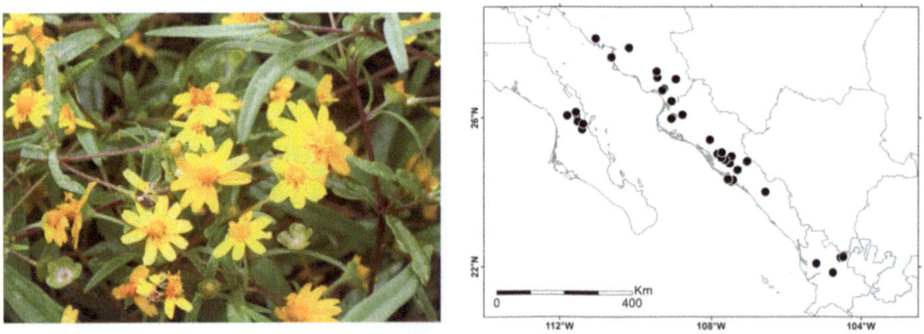

M. cupulatum

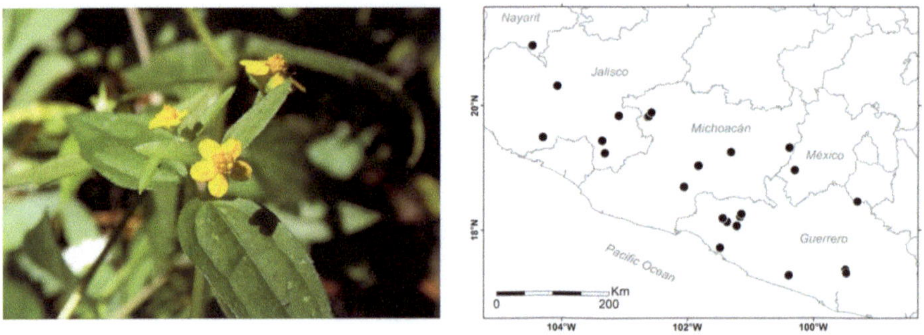

M. dicoelocarpum

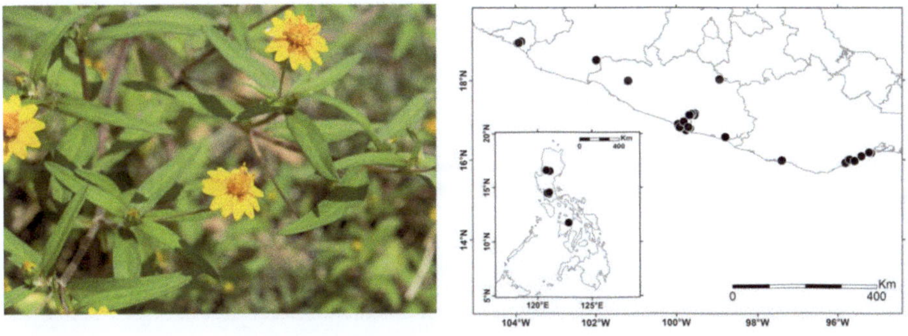

M. diffusum (inset is Philippine Islands)

M. divaricatum

M. glabribracteatum

M. glabrum

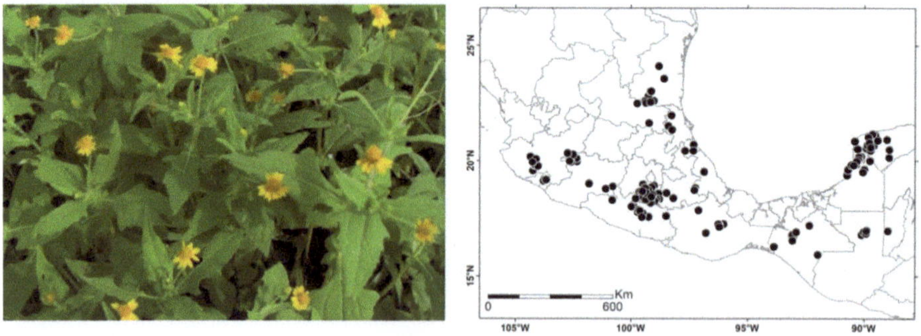

M. gracile

M. leucanthum

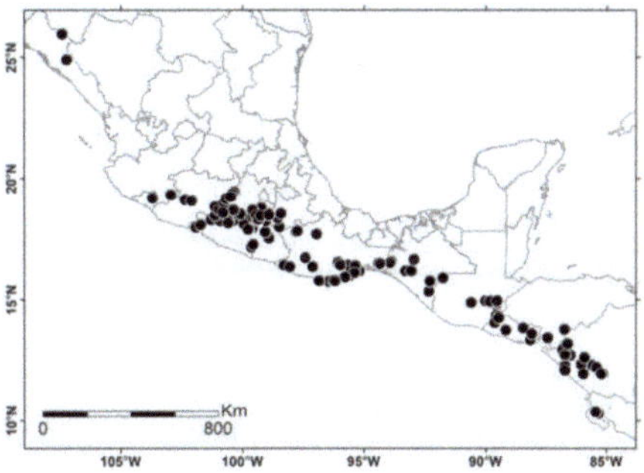

M. linearilobum

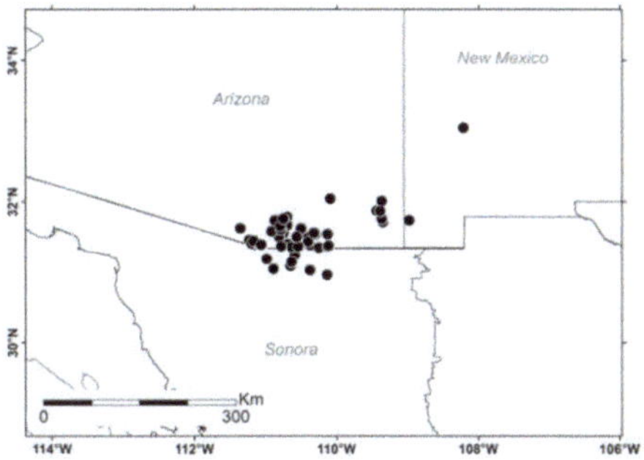

M. longicorne

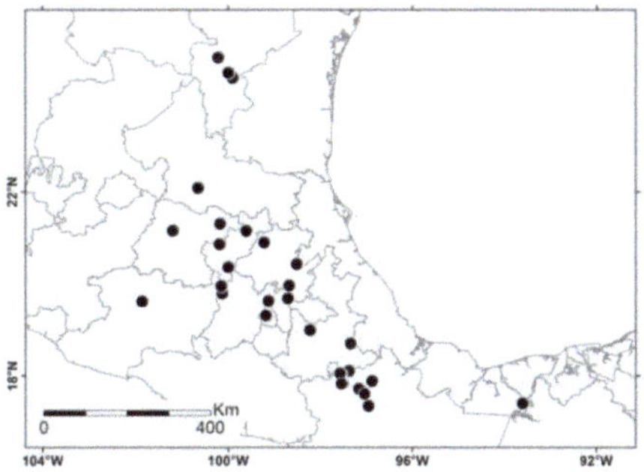

M. longifolium

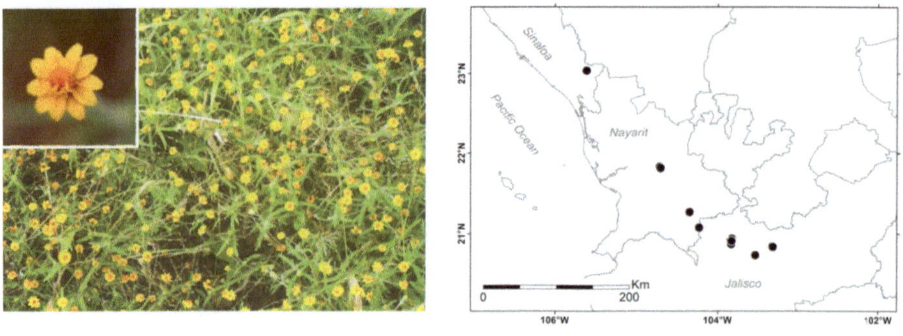

M. longipes

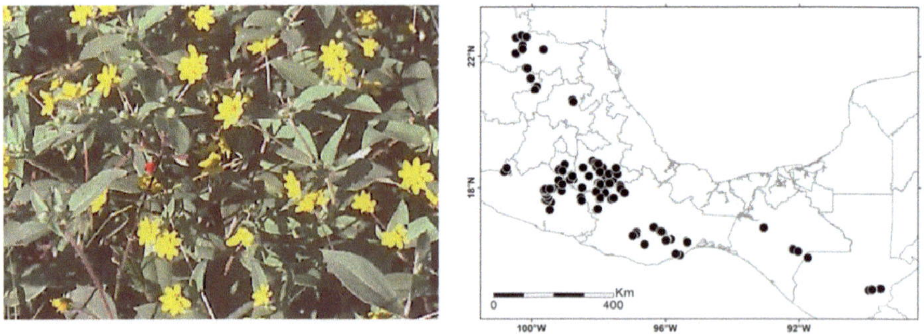

M. longipilum

M. mayfieldii

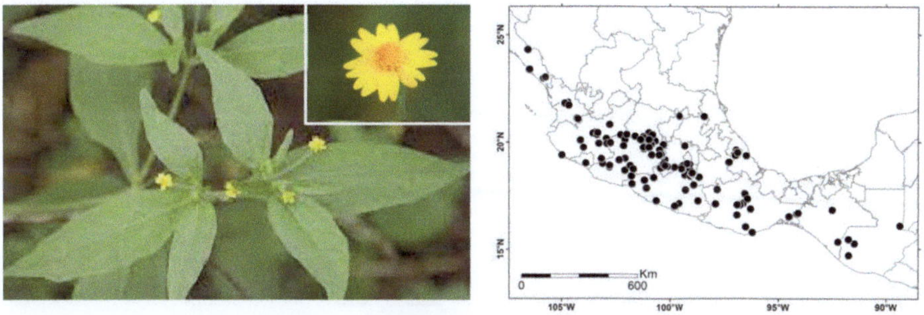

M. microcephalum

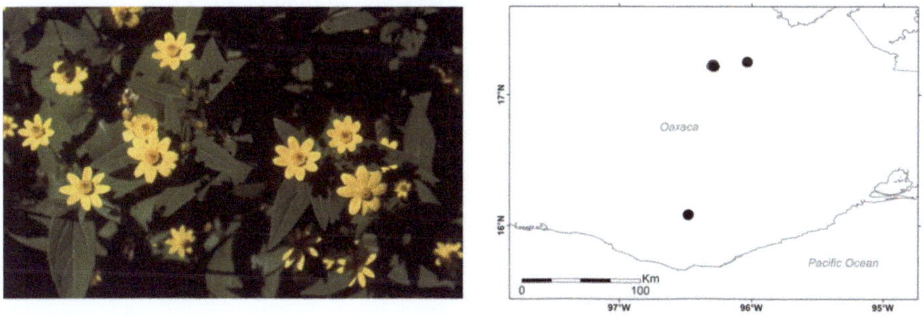

M. mimulifolium

M. moctezumum

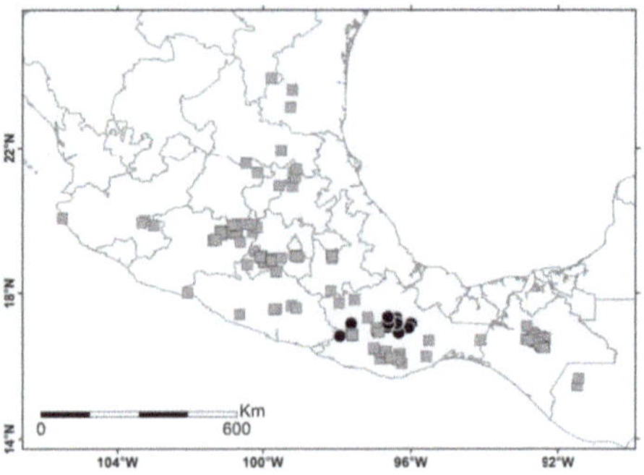

M. montanum (var. *montanum*, circles; var. *viridulum*, squares)

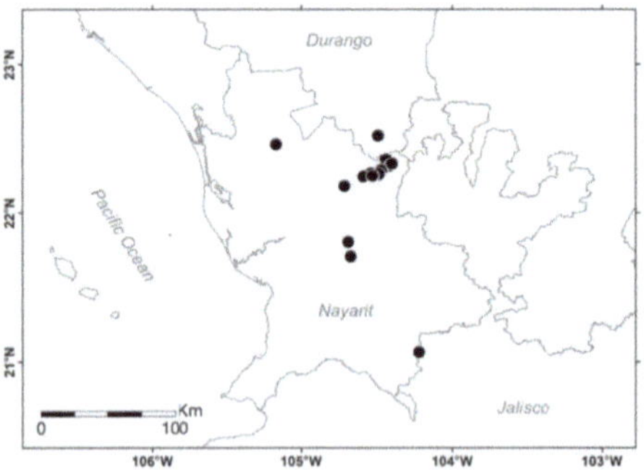

M. nayaritense

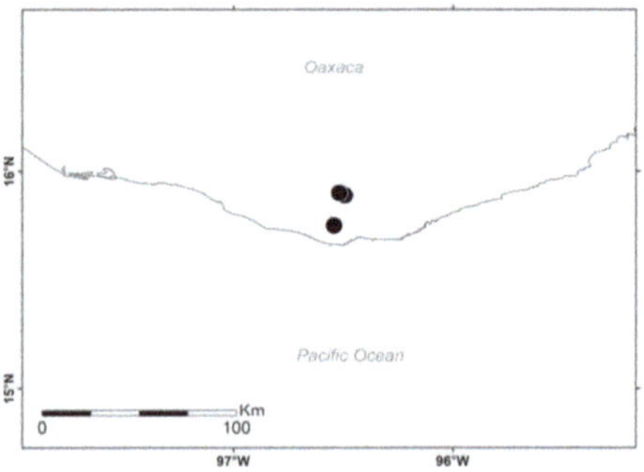

M. northingtonii

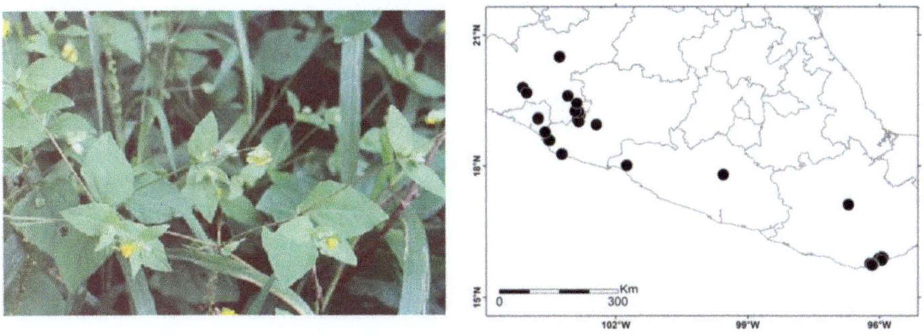

M. nutans

M. paniculatum

M. perfoliatum

M. pilosum

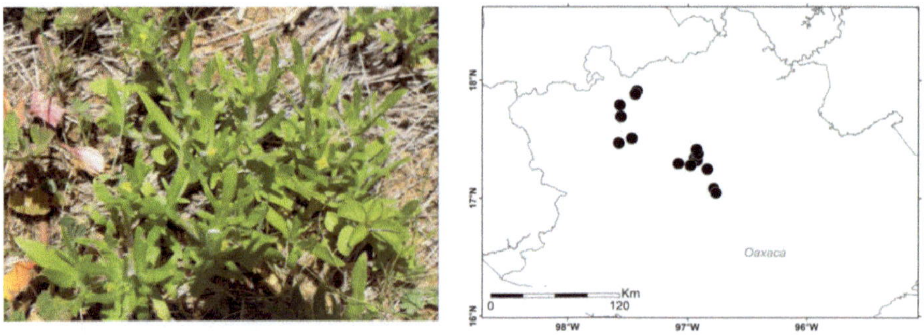

M. pringlei

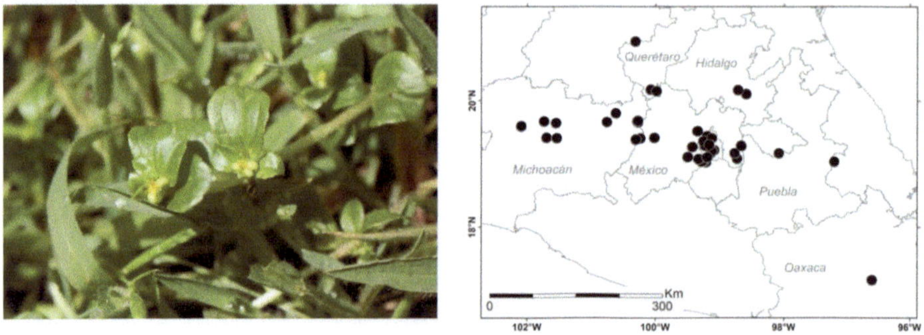

M. repens

M. rosei

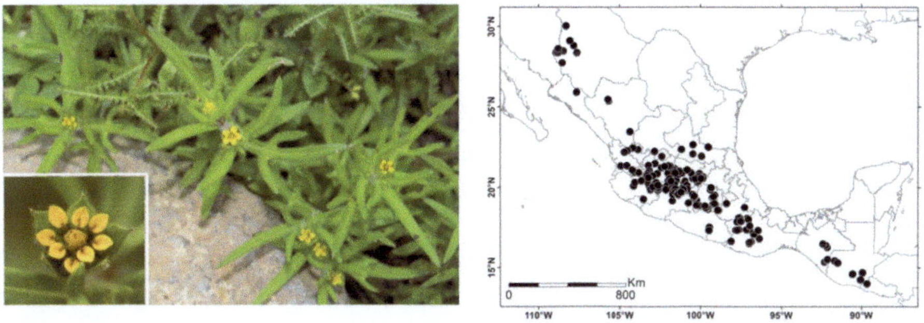

M. sericeum

M. sinaloense

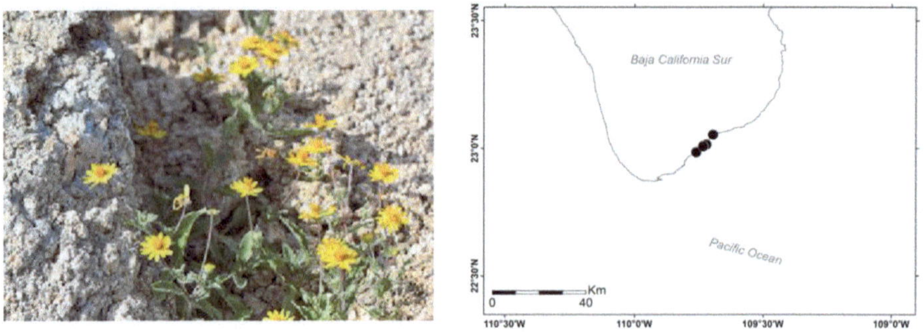

M. sinuatum

M. strigosum

M. tenellum

M. tepicense

Literature Cited

Achard, F., Eva, H. D., Stibig, H.-J., Mayaux, P., Gallego, J., Richards, T., & Malingreau, J.-P. (2002). Determination of deforestation rates of the world's humid tropical forests. *Science, 297*, 999–1002.

Adsersen, H. (1980). Revision of the Galápagos endemic genus *Lecocarpus* (Asteraceae). *Botanisk Tidsskrift, 75*, 63–76.

Ainouche, M. L., Fortune, P. M., Salmon, A., Parisod, C., Grandbastien, M.-A., Fukunaga, K., Ricou, M., & Misset, M.-T. (2009). Hybridization, polyploidy and invasion: Lessons from *Spartina* (Poaceae). *Biological Invasions, 11*, 1159–1173.

Anderson, E. (1948). Hybridization of the habitat. *Evolution, 2*, 1–9.

Anonymous. (1994). *Systematics agenda 2000: Charting the biosphere*. American Museum of Natural History and New York Botanical Garden: New York.

Avise, J. C. (2000). *Phylogeography: The history and formation of species*. Harvard University Press: Cambridge.

Avise, J. C. (2008). Phylogeography: Retrospect and prospect. *Journal of Biogeography, 36*, 3–15.

Baillon, H. E. (1882). *Histoire des Plantes* (Vol. 8, p. 231). Paris.

Baker, J. G. (1884). *Melampodium*. In C. F. P. von Martius (Ed.), *Flora Brasiliensis* (Vol. 6(3), pp. 159–162). Munich.

Baldwin, B. G. (2009). Heliantheae alliance. In V. A. Funk, A. Susanna, T. F. Stuessy, & R. J. Bayer (Eds.), *Systematics, evolution, and biogeography of Compositae* (pp. 689–711). International Association for Plant Taxonomy: Vienna.

Baldwin, B. G., Sanderson, M. J., Porter, J. M., Wojciechowski, M. F., Campbell, C. S., & Donoghue, M. J. (1995). The ITS region of nuclear ribosomal DNA: A valuable source of evidence on angiosperm taxonomy. *Annals of the Missouri Botanical Garden, 82*, 247–277.

Banks, J. 1781. *Reliquiae Houstounianae: seu plantarum in America meridionali a Gulielmo Houstoun M.D.R.S.S. Collectarum icones manu propria aere incisae; cum descriptionibus e schedis ejusdem in bibliotheca Josephi Banks, Baroneti, R.S.P. asservatis*. London.

Becerra, J. X. (2005). Timing the origin and expansion of the Mexican tropical dry forest. *Proceedings of the National Academy of Sciences of the United States of America, 102*, 10919–10923.

Bennetzen, J. L. (2005). Mechanisms of recent genome size variation in flowering plants. *Annals of Botany, 95*, 127–132.

Bennetzen, J. L., & Wang, H. (2014). The contributions of transposable elements to the structure, function, and evolution of plant genomes. *Annual Review of Plant Biology, 65*, 505–530.

Bentham, G. (1873). Notes on the classification, history, and geographical distribution of Compositae. *Journal of The Linnean Society of London, Botany, 13*, 335–577.

T. F. Stuessy et al., *Systematics, Ecology, and Chromosomal Evolution of* Melampodium *(Asteraceae)*, https://doi.org/10.1007/978-3-031-81995-7

Bentham, G., & Hooker, J. D. (1873). *Melampodium*. In *Genera plantarum* (Vol. 2, pp. 348–349). Reeve & Co.: London.

Bernardini, S., Osorio, M. S., & Tiezzi, A. (2018). Plants: An infinite source of molecules useful for pharmaceuticals. *Current Traditional Medicine, 4*, 157–165.

Blake, S. F. (1921). Revision of the genus *Acanthospermum*. *Contributions from the United States National Herbarium, 20*, 383–392. 1 plate.

Blake, S. F. (1922). Two new species of *Acanthospermum* from the Galápagos Islands. *Journal of the Washington Academy of Sciences, 12*, 200–205.

Blake, S. F. (1924). New American Asteraceae. *Contributions from the United States National Herbarium, 22*, 587–661.

Blanckaert, A., Bank, C., & Hermisson, J. (2020). The limits to parapatric speciation 3: Evolution of strong reproductive isolation in presence of gene flow despite limited ecological differentiation. *Philosophical Transactions of the Royal Society B, 375*, 20190532.

Blaxter, M. L. (2004). The promise of a DNA taxonomy. *Philosophical Transactions of the Royal Society B, 359*, 669–679.

Blöch, C. (2010). *Molecular phylogeny and chromosome evolution of the genus Melampodium L. (Millerieae, Asteraceae)*. Doctoral dissertation. University of Vienna: Vienna.

Blöch, C., Weiss-Schneeweiss, H., Schneeweiss, G. M., Barfuss, M. H. J., Rebernig, C. A., Villaseñor, J. L., & Stuessy, T. F. (2009). Molecular phylogenetic analyses of nuclear and plastid DNA sequences support dysploid and polyploid chromosome number changes and reticulate evolution in the diversification of *Melampodium* (Millerieae, Asteraceae). *Molecular Phylogenetics and Evolution, 53*, 220–233.

Bock, D. G., Cai, Z., Elphinstone, C., González-Segovia, E., Hirabayashi, K., Huang, K., Keais, G. L., Kim, A., Owens, G. L., & Rieseberg, L. H. (2023). Genomics of plant speciation. *Plant Communications, 4*, 100599.

Bohm, B. A. (1998). *Introduction to flavonoids*. Harwood: Amsterdam.

Bohm, B. A., & Stuessy, T. F. (1991). Flavonoid variation in *Melampodium*. *Biochemical Systematics and Ecology, 19*, 677–679.

Bohm, B. A., & Stuessy, T. F. (2001). *Flavonoids of the sunflower family* (Asteraceae). Springer: Vienna.

Borsch, T., Hernández-Ledesma, P., Berendsohn, W. G., Flores-Olvera, H., Ochoterena, H., Zuloaga, F. O., Mering, S., & Kilian, N. (2015). An integrative and dynamic approach for monographing species-rich plant groups—Building the global synthesis of the angiosperm order Caryophyllales. *Perspectives in Plant Ecology, Evolution and Systematics, 17*, 284–300.

Both, C., Bouwhuis, S., Lessells, C. M., & Visser, M. E. (2006). Climate change and population declines in a long-distance migratory bird. *Nature, 441*, 81–83.

Brochmann, C., Brysting, A. K., Alsos, I. G., Borgen, L., Grundt, H. H., Scheen, A. C., & Elven, R. (2004). Polyploidy in arctic plants. *Biological Journal of the Linnean Society, 82*, 521–536.

Brok, C. S., & Adsersen, H. (2007). Morphological variation among populations of *Lecocarpus* (Asteraceae) on the Galápagos Islands. *Botanical Journal of the Linnean Society, 154*, 523–544.

Brysting, A. K., Oxelman, B., Huber, K. T., Moulton, V., & Brochmann, C. (2007). Untangling complex histories of genome mergings in high polyploids. *Systematic Biology, 56*, 467–476.

Buggs, R. J. A., Soltis, P. S., Mavrodiev, E. V., Symonds, V. V., & Soltis, D. E. (2008). Does phylogenetic distance between parental genomes govern the success of polyploids? *Castanea, 73*, 74–93.

Buggs, R. J. A., Soltis, P. S., & Soltis, D. E. (2009). Does hybridization between divergent progenitors drive whole-genome duplication? *Molecular Ecology, 18*, 3334–3339.

Buggs, R. J. A., Soltis, P. S., & Soltis, D. E. (2011). Biosystematic relationships and the formation of polyploids. *Taxon, 60*, 324–332.

Büntgen, U., Piermattei, A., Krusic, P. J., Esper, J., Sparks, T., & Crivellaro, A. (2022). Plants in the UK flower a month earlier under recent warming. *Proceedings of the Royal Society B, 289*, 20212456. https://doi.org/10.1098/rspb.2021.2456

Burgman, M. A., & Fox, J. C. (2003). Bias in species range estimates from minimum convex polygons: Implications for conservation and options for improved planning. *Animal Conservation, 6*, 19–28.

Burnett, W. C., Jr., Jones, S. B., Jr., Mabry, T. J., & Padolina, W. G. (1974). Sesquiterpene lactones—insect feeding deterrents in *Vernonia*. *Biochemical Systematics and Ecology, 2*, 25–29.

Candolle, A. P. de (1836). *Melampodium*. In *Prodromus systematis naturalis regni vegetabilis, pars quinta* (pp. 517–521). Paris.

Carlquist, S., & Eckhart, V. M. (1982). Wood anatomy of *Darwiniothamnus*, *Lecocarpus*, and *Macraea* (Asteraceae). *Aliso, 10*, 291–300.

Carr, G. D., King, R. M., Powell, A. M., & Robinson, H. (1999). Chromosome numbers in Compositae. XVIII. *American Journal of Botany, 86*, 1003–1013.

Cassini, A. H. G. (1829). *Melampodium*, *Zarabellia* and *Alcina*. In G. Cuvier (Ed.), *Dictionnaire des Sciences Naturelles* (Vol. 59, 2nd ed., pp. 237–244). Paris.

Castoe, T. A., Spencer, C. L., & Parkinson, C. L. (2007). Phylogeographic structure and historical demography of the western diamondback rattlesnake (*Crotalus atrox*): A perspective on North American desert biogeography. *Molecular Phylogenetics and Evolution, 42*, 193–212.

Chapman, M. A., & Burke, J. M. (2007). Genetic divergence and hybrid speciation. *Evolution, 61*, 1773–1780.

Cockerell, T. D. A. (1905). The name *Melampodium*. *Torreya, 5*, 70.

Coleman, J. R. (1970). Additional chromosome numbers in Brazilian Compositae. *Rhodora, 72*, 94–99.

Costanza, R., d'Arge, R., de Groot, R., Farber, S., Grasso, M., Hannon, B., Limburg, K., Naeem, S., O'Neill, R. V., Paruelo, J., Raskin, R. G., Sutton, P., & van den Belt, M. (1997). The value of the world's ecosystem services and natural capital. *Nature, 387*, 253–260.

Coyne, J. A., & Orr, H. A. (2004). *Speciation*. Sinauer: Sunderland, Massachusetts.

Crawford, D. J. (2014). Primary or divergent speciation. In T. F. Stuessy, D. J. Crawford, D. E. Soltis, & P. S. Soltis (Eds.), *Plant systematics: The origin, interpretation, and ordering of plant biodiversity* (pp. 111–124). Koeltz Scientific Books: Königstein.

Crawford, D. J., Kelly, J. K., & Anderson, G. J. (2023). Reproductive biology of Asteraceae on oceanic islands. *The Botanical Review, 89*. https://doi.org/10.1007/s12229-023-09295-9

Cronquist, A. (1970). New combinations in the Compositae of the Galápagos Islands. *Madroño, 20*, 255–256.

Cronquist, A. (1971a). *Acanthospermum*. In I. L. Wiggins & D. M. Porter (Eds.), *Flora of the Galápagos Islands* (pp. 307–309). Stanford University Press: Stanford.

Cronquist, A. (1971b). *Lecocarpus*. In I. L. Wiggins & D. M. Porter (Eds.), *Flora of the Galápagos Islands* (pp. 342–344). Stanford University Press: Stanford.

Daly, M., Herendeen, P. S., Guralnick, R. P., Westneat, M. W., & McDade, L. (2012). Systematics agenda 2020: The mission evolves. *Systematic Biology, 61*, 549–552.

Dandy, J. E. (Rev. and Ed.). (1958). *The Sloane herbarium: An annotated list of the* horti sicci *composing it; with biographical accounts of the principal contributors*. British Museum: London.

de Vos, J. M., Augustijnen, H., Bätscher, L., & Lucek, K. (2020). Speciation through chromosomal fusion and fission in Lepidoptera. *Philosophical Transactions of the Royal Society B, 375*, 20190539. https://doi.org/10.1098/rstb.2019.0539

Deanna, R., Acosta, M. C., Scaldaferro, M., & Chiarini, F. (2022). Chromosome evolution in the family Solanaceae. *Frontiers in Plant Science, 12*, 787590. https://doi.org/10.3389/fpls.2021.787590

Decaisne, M. J. (1846). Atlas de botanique. In D. Petit-Thouars (Ed.), *Voyage autour du monde sur la frégate La Vénus*. Paris. [*Lecocarpus pinnatifidus*, t. 14.].

Dobzhansky, T. (1937). *Genetics and the evolution of species*. Columbia University Press: New York.

Dodsworth, S., Jang, T.-S., Struebig, M., Chase, M. W., Weiss-Schneeweiss, H., & Leitch, A. R. (2016). Genome-wide repeat dynamics reflect phylogenetic distance in closely related allotetraploid *Nicotiana* (Solanaceae). *Plant Systematics and Evolution, 303*, 1013–1020.

Doležel, J., Greilhuber, J., & Suda, J. (2007). Estimation of nuclear DNA content in plants using flow cytometry. *Nature Protocols, 2*, 2233–2244.

Eilam, T., Anikster, Y., Millet, E., Manisterski, J., & Feldman, M. (2009). Genome size and natural and synthetic autopolyploids and in a natural segmental allopolyploid of several Triticeae species. *Genome, 52*, 275–285.

Eliasson, U. (1971). Studies in Galápagos plants X. The genus *Lecocarpus* Decaisne. *Svensk Botanisk Tidskrift, 65*, 245–277.

ESRI. (2013). *ArcGis*. Environmental Systems Research Institute: Redlands.

Faegri, K., & van der Pijl, L. (1971). *The principles of pollination ecology* (2nd ed.). Pergamon Press: Oxford.

Farkic, J., Isailovic, G., & Taylor, S. (2021). Forest bathing as a mindful tourism practice. *Annals of Tourism Research Empirical Insights, 2*. https://doi.org/10.1016/j.annale.2021.100028

Fehlberg, S. D., & Ranker, T. A. (2009). Evolutionary history and phylogeography of *Encelia farinosa* (Asteraceae) from the Sonoran, Mojave, and Peninsular Deserts. *Molecular Phylogenetics and Evolution, 50*, 326–335.

Ferrari, L., Orozco-Esquivel, T., Manea, V., & Manea, M. (2012). The dynamic history of the trans-Mexican Volcanic Belt and the Mexico subduction zone. *Tectonophysics, 522-523*, 122–149.

Ferrari, L., Valencia-Moreno, M., & Bryan, S. (2007). Magmatism and tectonics of the Sierra Madre Occidental and its relation with the evolution of the western margin of North America. *Geological Society of America Special Papers, 422*, 1–39.

Ferrusquía-Villafranca, I. (2007). Ensayo sobre la caracterización y significación biológica. In I. Luna, J. J. Morrone, & D. Espinoza (Eds.), *Biodiversidad de la Faja Volcánica Transmexicana* (pp. 7–23). Universidad Nacional Autónoma de México: Ciudad de México.

Fleischmann, A., Michael, T. P., Rivadavia, F., Sousa, A., Wang, W., Temsch, E. M., Greilhuber, J., Müller, K. F., & Heubl, G. (2014). Evolution of genome size and chromosome number in the carnivorous plant genus *Genlisea* (Lentibulariaceae), with a new estimate of the minimum genome size in angiosperms. *Annals of Botany, 114*, 1651–1663.

Funk, V. A., Bayer, R. J., Keeley, S., Chan, R., Watson, L., Gemeinholzer, B., Schilling, E., Panero, J. L., Baldwin, B. G., Garcia-Jacas, N., Susanna, A., & Jansen, R. K. (2005). Everywhere but Antarctica: Using a supertree to understand the diversity and distribution of the Compositae. In I. Friis & H. Balslev (Eds.), *Plant diversity and complexity patterns: Local, regional and global dimensions* (pp. 343–374). The Royal Danish Academy of Sciences and Letters: Copenhagen. [*Biol. Skrift*. 55.].

Funk, V. A., Susanna, A., Stuessy, T. F., & Bayer, R. J. (Eds.). (2009). *Systematics, evolution, and biogeography of Compositae*. International Association for Plant Taxonomy: Vienna.

Gardner, R. C. (1977). Observations on tetramerous disc florets in the Compositae. *Rhodora, 79*, 139–146.

Garrido-Ramos, M. A. (2015). Satellite DNA in plants: More than just rubbish. *Cytogenetic and Genome Research, 146*, 153–170.

Gómez-Tuena, A., Orozco-Esquivel, M. T., & Ferrari, L. (2005). Petrogénesis ígnea de la Faja Volcánica Transmexicana. *Boletín de la Sociedad Geológica Mexicana, 57*, 227–283.

Grace, O. M., Pérez-Escobar, O. A., Lucas, E. J., Vorontsova, M. S., Lewis, G. P., Walker, B. E., Lohmann, L. G., Knapp, S., Wilkie, P., Sarkinen, T., Darbyshire, I., Lughadha, E. N., Monro, A., Woudstra, Y., Demissew, S., Muasya, A. M., Díaz, S., Baker, W. J., & Antonelli, A. (2021). Botanical monography in the Anthropocene. *Trends in Plant Science, 26*, 433–441.

Grant, V. (1981). *Plant speciation* (2nd ed.). Columbia University Press: New York.

Gray, A. (1884). *Synoptical flora of North America* (Vol. 1(2), pp. 238–239). Ivison, Blakeman, Taylor, and Co.: New York.

Harborne, J. B., & Turner, B. L. (1984). *Plant chemosystematics*. Academic Press: London.

Hartman, R. L., & Stuessy, T. F. (1983). Revision of *Otopappus* (Compositae, Heliantheae). *Systematic Botany, 8*, 185–210.

Hebert, P. D. N., Cywinska, A., Ball, S. L., & deWaard, J. R. (2003). Biological identifications through DNA barcodes. *Proceedings of the Royal Society B, 270*, 313–321.

Hillis, D. M., Moritz, C., & Mable, B. K. (Eds.). (1996). *Molecular systematics* (2nd ed.). Sinauer: Sunderland, Massachusetts.

Hoffmann, O. (1890). *Melampodium*. In A. Engler & E. Prantl (Eds.), *Die natürlichen Pflanzenfamilien* (Vol. 4(5), p. 218). Engelmann: Leipzig.

Holland, R. B., Huber, K. T., Moulton, V., & Lockhart, P. J. (2004). Using consensus networks to visualize contradictory evidence for species phylogeny. *Molecular Biology and Evolution, 21*, 1459–1461.

Holmgren, C. A., Norris, J., & Betancourt, J. L. (2007). Inference about winter temperature and summer rains from the late quaternary record of C_4 perennial grasses and C_3 desert shrubs in the northern Chihuahuan Desert. *Journal of Quaternary Science, 22*, 141–161.

Humboldt, F. H. A., Bonpland, A. J., & Kunth, C. S. (1818) [folio]. *Melampodium*. In *Nova genera et species plantarum* (Vol. 4, pp. 272–274). Paris.

Hunter, K. L., Betancourt, J. L., Riddle, B. R., Van Devender, T. R., Cole, K. L., & Spaulding, W. G. (2001). Ploidy race distributions since the last glacial maximum in the North American desert shrub, *Larrea tridentata*. *Global Ecology and Biogeography, 10*, 521–533.

IUCN. (2022). *Red list of threatened species*. Summary statistics; Table 1a. www.iucnredlist.org/resources/.summary-statistics#Summary%20Tables

Jackson, R. C. (1957). Documented chromosome numbers of plants. *Madroño, 14*, 111.

Jaeger, J. R., Riddle, B. R., & Bradford, D. F. (2005). Cryptic Neogene vicariance and quaternary dispersal of the red-spotted toad (*Bufo punctatus*): Insights on the evolution of North American warm desert biotas. *Molecular Ecology, 14*, 3033–3048.

James, S. H. (1979). Chromosome numbers and genetic systems in the trigger plants of western Australia (*Stylidium*; Stylidiaceae). *Australian Journal of Botany, 27*, 17–25.

Jang, T.-S., Parker, J. S., Emadzade, K., Temsch, E. M., Leitch, A. R., & Weiss-Schneeweiss, H. (2018). Multiple origins and nested cycles of hybridization result in high tetraploid diversity in the monocot *Prospero*. *Frontiers in Plant Science, 9*, 3433.

Jarvis, C. (2007). *Order out of chaos: Linnaean plant names and their types*. Linnean Society of London and Natural History Museum: London.

Johnson, M. D., Freeland, J. R., Parducci, L., Evans, D. M., Meyer, R. S., Molano-Flores, B., & Davis, M. A. (2022). Environmental DNA as an emerging tool in botanical research. *American Journal of Botany, 110*. https://doi.org/10.1002/ajb2.16120

Jones, K. (1978). Aspects of chromosome evolution in higher plants. *Advances in Botanical Research, 6*, 119–194.

Kareiva, P., Watts, S., McDonald, R., & Boucher, T. (2007). Domesticated nature: Shaping landscapes and ecosystems for human welfare. *Science, 316*, 1866–1869.

Kawecki, T. J. (2004). Genetic theories of sympatric speciation. In U. Dieckmann, M. Doebeli, J. A. J. Metz, & D. Tautz (Eds.), *Adaptive speciation* (pp. 36–53). Cambridge University Press: Cambridge.

Keil, D. J., & Stuessy, T. F. (1975). Chromosome counts of Compositae from the United States, Mexico, and Guatemala. *Rhodora, 77*, 171–195.

Keil, D. J., & Stuessy, T. F. (1977). Chromosome counts of Compositae from Mexico and the United States. *American Journal of Botany, 64*, 791–798.

Keil, D. J., & Stuessy, T. F. (1981). Systematics of *Isocarpha* (Compositae: Eupatorieae). *Systematic Botany, 6*, 258–287.

Kelley, L. J., Renny-Byfield, S., Pellicer, J., Macas, J., Novák, P., Neumann, P., Lysak, M. A., Day, P. D., Berger, M., Fay, M. F., Nichols, R. A., Leitch, A. R., & Leitch, I. J. (2015). Analysis of the giant genomes of *Fritillaria* (Liliaceae) indicates that a lack of DNA removal characterizes extreme expansions in genome size. *The New Phytologist, 208*, 596–607.

Kim, S. T., Sultan, S. E., & Donoghue, M. J. (2008). Allopolyploid speciation in *Persicaria* (Polygonaceae): Insights from a low-copy nuclear region. *Proceedings of the National Academy of Sciences of the United States of America, 105*, 12370–12375.

King, M. (1993). *Species evolution: The role of chromosome change*. Cambridge University Press: Cambridge.

Kolář, F., Čertner, M., Suda, J., Schöneswetter, P., & Husband, B. C. (2017). Mixed-ploidy species: Progress and opportunities in polyploid research. *Trends in Plant Science, 22*, 1041–1055.

Kron, P., Suda, J., & Husband, B. C. (2007). Applications of flow cytometry to evolutionary and population biology. *Annual Review of Ecology, Evolution, and Systematics, 38*, 847–876.

Lee, J., Park, B.-J., Tsunetsugu, Y., Ohira, T., Kagawa, T., & Miyazaki, Y. (2011). Effect of forest bathing on physiological and psychological responses in young Japanese male subjects. *Public Health, 125*, 93–100.

Leitch, A. R., & Leitch, I. J. (2008). Genome plasticity and the diversity of polyploid plants. *Science, 320*, 481–483.
Leitch, I. J., Hanson, L., Lim, K. Y., Kovařík, A., Chase, M. W., Clarkson, J. J., & Leitch, A. R. (2008). The ups and downs of genome size evolution in polyploid species of *Nicotiana* (Solanaceae). *Annals of Botany, 101*, 805–814.
Leopold, A. S. (1950). Vegetation zones of Mexico. *Ecology, 31*, 507–518.
Lessing, C. F. (1830). Melampodieae, p. 149. In: Schlechtendal, D. and Chamisso, A. de, Plantarum Mexicanarum a cel. viris Schiede et Deppe collectarum. *Linnaea, 5*, 72–174, 206–236, 554–625.
Lessing, C. F. (1832). *Synopsis generum Compositarum*. Berlin.
Levin, D. A. (2000). *The origin, expansion, and demise of plant species*. Oxford University Press: Oxford.
Levin, D. A. (2002). *The role of chromosomal change in plant evolution*. Oxford University Press: Oxford.
Lewis, H. (1966). Speciation in flowering plants. *Science, 152*, 167–172.
Lewis, W. H. (1962). Aneusomaty in aneuploid populations of *Claytonia virginica*. *American Journal of Botany, 49*, 918–928.
Lewis, W. H. (1970). Extreme instability of chromosome number in *Claytonia virginica*. *Taxon, 19*, 180–182.
Li, Q. (2010). Effect of forest bathing trips on human immune function. *Environmental Health and Preventive Medicine, 15*, 9–17.
Linder, C. R., & Rieseberg, L. H. (2004). Reconstructing patterns of reticulate evolution in plants. *American Journal of Botany, 91*, 1700–1708.
Linnaeus, C. (1737a). *Hortus Cliffortianus*. Amsterdam.
Linnaeus, C. (1737b). *Critica botanica*. Leiden.
Linnaeus, C. (1753). *Species plantarum*. Stockholm.
Linnaeus, C. (1754). *Genera plantarum* (5th ed.). Stockholm.
Linnaeus, C. *filius*. (1781). Supplementum plantarum. Braunschweig.
Löve, A., & Löve, D. (1964). The biological species concept and its evolutionary structure. *Taxon, 13*, 33–45.
Löve, A., Löve, D., & Raymond, M. (1957). Cytotaxonomy of *Carex* section *Capillares*. *Canadian Journal of Botany, 35*, 715–761.
Mabry, T. J., Markham, K. R., & Thomas, M. B. (1970). *The systematic identification of flavonoids*. Springer: New York.
Macas, J., Novák, P., Pellicer, J., Čížková, J., Koblížková, A., Neumann, P., Fuková, I., Doležel, J., Kelly, L. J., & Leitch, I. J. (2015). In depth characterization of repetitive DNA in 23 plant genomes reveals sources of genome size variation in the legume tribe Fabeae. *PLoS One, 10*, e0143424. https://doi.org/10.1371/journal.pone.0143424
Maddison, W. P., & Maddison, D. R. (2021). *Mesquite: A modular system for evolutionary analysis. Version, 3.70*. http://mesquiteproject.org
Manilal, K. S. (1973). Morphology of the capitulum of *Melampodium divaricatum* (Rich.) DC. *Current Science, 42*, 578–580.
Manilal, K. S. (1975). A compound capitulum in *Melampodium* L. (Compositae). *Botanical Journal of the Linnean Society, 70*, 71–75.
Marhold, K., & Stuessy, T. (Eds.), in collaboration with Agababian, M., Agosti, D., Alford, M. H., Crespo, A., Crisci, J. V., Dorr, L. J., Ferencová, Z., Frodin, D., Geltman, D. V., Kilian, N., Linder, H. P., Lohmann, L. G., Oberprieler, C., Penev, L., Smith, G. F., Thomas, W., Tulig, M., Turland, N., & Zhang, X.-C. (2013). The future of botanical monography: Report from an international workshop, 12–16 March 2012, Smolenice, Slovak Republic. *Taxon, 62*, 4–20.
Markham, K. R. (1982). *Techniques of flavonoid identification*. Academic Press: London.
Masterson, J. (1994). Stomatal size in fossil plants: Evidence for polyploidy in majority of angiosperms. *Science, 264*, 421–423.

Mastretta-Yanes, A., Moreno-Letelier, A., Piñero, D., Jorgensen, T. H., & Emerson, B. C. (2015). Biodiversity in the Mexican highlands and the interaction of geology, geography and climate within the Trans-Mexican Volcanic Belt. *Journal of Biogeography, 42*, 1586–1600.
Mayden, R. L. (1997). A hierarchy of species concepts: The denouement in the saga of the species problem. In M. F. Claridge, H. A. Dawah, & M. R. Wilson (Eds.), *Species: The units of biodiversity* (pp. 381–424). Chapman and Hall: London.
Mayr, E. (1954). Change of genetic environment and evolution. In J. Huxley, A. C. Hardy, & E. B. Ford (Eds.), *Evolution as a process* (pp. 157–180). Allen and Unwin: London.
Mayr, E. (1957). Species concepts and definitions. In E. Mayr (Ed.), *The species problem* (pp. 1–22). American Association for the Advancement of Science: Washington, D.C.
McCann, J., Jang, T.-S., Macas, J., Schneeweiss, G. M., Matzke, N. J., Novak, P., Stuessy, T. F., Villaseñor, J. L., & Weiss-Schneeweiss, H. (2018). Dating the species network: Allopolyploidy and repetitive DNA evolution in American daisies (*Melampodium* sect. *Melampodium*, Asteraceae). *Systematic Biology, 67*, 1010–1024.
McCann, J., Macas, J., Novak, P., Stuessy, T. F., Villaseñor, J. L., & Weiss-Schneeweiss, H. (2020). Differential genome size and repetitive DNA evolution in diploid species of *Melampodium* sect. *Melampodium* (Asteraceae). *Frontiers in Plant Science, 11*, 362. https://doi.org/10.3389/fpls.2020.00362
McCann, J., Schneeweiss, G. M., Stuessy, T. F., Villaseñor, J. L., & Weiss-Schneweiss, H. (2016). The impact of reconstruction methods, phylogenetic uncertainty and branch lengths on inference of chromosome number evolution in American daisies (*Melampodium*, Asteraceae). *PLoS One, 11*, e0162299. https://doi.org/10.1371/journal.pone.0162299
Melters, D. P., Bradnam, K. R., Young, H. A., Telis, N., May, M. R., Ruby, J. G., Sebra, R., Peluso, P., Eid, J., Rank, D., Garcia, J. F., DeRisi, J. L., Smith, T., Tobias, C., Ross-Ibarra, J., Korf, I., & Chan, S. W. L. (2013). Comparative analysis of tandem repeats from hundreds of species reveals unique insights into centromere evolution. *Genome Biology, 14*, Ra10. https://doi.org/10.1186/gb-2013-14-1-r10
Merrill, E. D. (1954). The botany of Cook's voyages. *Chronica Botanica, 14*, 161–384.
Meyer, L., Diniz-Filho, J. A., & Lohmann, L. G. (2017). A comparison of hull methods for estimating species ranges and richness maps. *Plant Ecology & Diversity, 10*, 389–401.
Miller, J. S. (1996). Collecting methodologies for plant samples for pharmaceutical research. In T. F. Stuessy & S. H. Sohmer (Eds.), *Sampling the green world: Innovative concepts of collection, preservation, and storage of plant diversity* (pp. 74–87). Columbia University Press: New York.
Miller, P. (1731). *The Gardner's dictionary*. London.
Miller, S. E. (2015). DNA barcoding in floral and faunal research. In M. F. Watson, C. H. C. Lyal, & C. A. Pendry (Eds.), *Descriptive taxonomy: The foundation of biodiversity research* (pp. 296–311). Cambridge University Press: Cambridge.
Miranda, F., & Hernández, X. E. (1963). Los tipos de vegetación de México y su clasificación. *Boletín de la Sociedad Botánica de México, 28*, 29–179.
Mitchard, E. T. A. (2018). The tropical forest carbon cycle and climate change. *Nature, 559*, 527–534.
Mora, C., Tittensor, D. P., Adl, S., Simpson, A. G. B., & Worm, B. (2011). How many species are there on earth and in the ocean? *PLoS Biology, 9*(8), e1001127. https://doi.org/10.1371/journal.pbio.1001127
Moran-Zenteno, D. (1994). *The geology of the Mexican Republic*. American Association of Petroleum Geologists: Tulsa. [Report No. 39].
Naeem, S., Duffy, J. E., & Zavaleta, E. (2012). The functions of biological diversity in an age of extinction. *Science, 336*, 1401–1406.
Negodi, G. (1938). Contributo alla cariologia ed alla morfologia di alcuni *Melampodium* (Compositae, Tubuliflorae-Heliantheae-Melampodinae). *Annali di Botanica (Rome), 21*, 495–502.
Nosil, P. (2008). Speciation with gene flow could be common. *Molecular Ecology, 17*, 2103–2106.
Nosil, P. (2012). *Ecological speciation*. Oxford University Press: Oxford.

Noss, R. F. (1990). Indicators for monitoring biodiversity: A hierarchical approach. *Conservation Biology, 4*, 355–364.

Novák, P., Hřibová, E., Neumann, P., Koblížková, A., Doležel, J., & Macas, J. (2014). Genome-wide analysis of repeat diversity across the family Musaceae. *PLoS One, 9*, e98918. https://doi.org/10.1371/journal.pone.0098918

Novák, P., Neumann, P., & Macas, J. (2020). Global analysis of repetitive DNA from unassembled sequence reads using RepeatExplorer2. *Nature Protocols, 15*, 3745–3776.

Novák, P., Neumann, P., Pech, J., Steinhaisl, J., & Macas, J. (2013). RepeatExplorer: A galaxy-based web server for genome-wide characterization of eukaryotic repetitive elements from next-generation sequence reads. *Bioinformatics, 29*, 792–793.

Olson, D. M., Dinerstein, E., Wikramanayake, E. D., Burgess, N. D., Powell, G. V. N., Underwood, E. C., D'Amico, J. A., Itoua, I., Strand, H. E., Morrison, J. C., Loucks, C. J., Allnutt, T. F., Ricketts, T. H., Kura, Y., Lamoreux, J. F., Wettengel, W. W., Hedao, P., & Kassem, K. R. (2001). Terrestrial ecoregions of the world: A new map of life on earth. *Bioscience, 51*, 933–938.

Otto, S. P., & Whitton, J. (2000). Polyploid incidence and evolution. *Annual Review of Genetics, 34*, 401–437.

Owens, G. L., Huang, K., Todesco, M., & Rieseberg, L. H. (2023). Re-evaluating homoploid reticulate evolution in *Helianthus* sunflowers. *Molecular Biology and Evolution, 40*, msad013.

Ozkan, H., Tuna, M., & Galbraith, D. W. (2006). No DNA loss in autotetraploids of *Arabidopsis thaliana*. *Plant Breeding, 125*, 288–291.

Palmer, J. D. (1987). Chloroplast DNA evolution and biosystematic uses of chloroplast DNA variation. *American Naturalist, 130*(suppl), S6–S29.

Pandey, A. K., & Dhakal, M. R. (2001). Phytomelanin in Compositae. *Current Science, 80*, 933–940.

Pandey, A. K., Stuessy, T. F., & Mathur, R. R. (2015). Phytomelanin and systematics of the Heliantheae alliance (Compositae). *Plant Diversity and Evolution, 131*, 145–165.

Pandey, A. K., Wilcox, L. W., Sack, F. D., & Stuessy, T. F. (1989). Development of the phytomelanin layer in fruits of *Ageratum conyzoides* (Compositae). *American Journal of Botany, 76*, 739–746.

Panero, J. L. (2007). Millerieae Lindl. (1829). In J. W. Kadereit & C. Jeffrey (Eds.), *Flowering plants, Eudicots, Asterales*. (Kubitzki, K. (gen. Ed.), *The families and genera of vascular plants*) (Vol. VIII, pp. 477–492). Springer: Berlin.

Panero, J. L., & Funk, V. A. (2002). Toward a phylogenetic subfamilial classification for the Compositae (Asteraceae). *Proceedings of the Biological Society of Washington, 115*, 909–922.

Parisod, D., Holderegger, R., & Brochmann, C. (2010). Evolutionary consequences of autopolyploidy. *New Phytologist, 186*, 5–17.

Pauly, D., & Maclean, J. (2003). *In a perfect ocean: The state of fisheries and ecosystems in the North Atlantic Ocean*. Island Press: Washington, D.C.

Paun, O., Fay, M., Forest, F., & Chase, M. W. (2009). Hybrid speciation in angiosperms: Parental divergence drives ploidy. *The New Phytologist, 182*, 507–518.

Pavlinov, I. Y. (2023). *The species problem: A conceptual history*. CRC Press: Boca Raton.

Pellicer, J., Fay, M. F., & Leitch, I. J. (2010). The largest eukaryotic genome of them all? *Botanical Journal of the Linnean Society, 164*, 10–15.

Pellicer, J., Hidalgo, O., Dodsworth, S., & Leitch, I. J. (2018). Genome size diversity and its impact on the evolution of land plants. *Genes, 9*, 88. https://doi.org/10.3390/genes9020088

Perrings, C., Dehnen-Schmutz, K., Touza, J., & Williamson, M. (2005). How to manage biological invasions under globalization. *Trends in Ecology & Evolution, 20*, 212–215.

Pesman, M. W. (1962). *Meet flora Mexicana: An easy way to recognize some of the more frequently met plants of Mexico as seen from the main highways*. Dale S. King: Globe, Arizona.

Philipp, M., Hansen, L. B., Adsersen, H., & Siegismund, H. R. (2004). Reproductive ecology of the endemic *Lecocarpus pinnatifidus* (Asteraceae) in an isolated population in the Galápagos Islands. *Botanical Journal of the Linnean Society, 146*, 171–180.

Picek, L., Šulc, M., Patel, Y., & Matas, J. (2022). Plant recognition by AI: Deep neural nets, transformers, and kNN in deep embeddings. *Frontiers in Plant Science, 13*, 787527.

Pitman, N. C. A., & Jørgensen, P. M. (2002). Estimating the size of the world's threatened flora. *Science, 298*, 989.

Potter, S., Bragg, J. G., Blom, M. P. K., Deakin, J. E., Kirkpatrick, M., Eldridge, M. D. B., & Moritz, C. (2017). Chromosomal speciation in the genomics era: Disentangling phylogenetic evolution of rock-wallabies. *Frontiers in Genetics, 8*, 10. https://doi.org/10.3389/fgene.2017.00010

Potts, S. G., Biesmeijer, J. C., Kremen, C., Neumann, P., Schweiger, O., & Kunin, W. E. (2010). Global pollinator declines: Trends, impacts and drivers. *Trends in Ecology & Evolution, 25*, 345–353.

Powell, A. M., & Sikes, S. (1970). Chromosome numbers of some Chihuahuan Desert Compositae. *The Southwestern Naturalist, 15*, 175–186.

Powell, K. I., Chase, J. M., & Knight, T. M. (2011). A synthesis of plant invasion effects on biodiversity across spatial scales. *American Journal of Botany, 98*, 539–548.

R Core Team. (2024). *R: A language and environment for statistical computing*. R Foundation for Statistical Computing: Vienna.

Rauscher, J. T. (2002). Molecular phylogenetics of the *Espeletia* complex (Asteraceae): Evidence from nrDNA ITS sequences on the closest relatives of an Andean adaptive radiation. *American Journal of Botany, 89*, 1074–1084.

RBG Kew. (2016). *The state of the world's plants—2016*. Royal Botanic Gardens: Kew.

Rebernig, C. A., Schneeweiss, G. M., Bardy, K. E., Schönswetter, P., Villaseñor, J. L., Obermayer, R., Stuessy, T. F., & Weiss-Schneeweiss, H. (2010a). Multiple Pleistocene refugia and Holocene range expansion of an abundant southwestern American desert plant species (*Melampodium leucanthum*, Asteraceae). *Molecular Ecology, 19*, 3421–3443.

Rebernig, C. A., Weiss-Schneeweiss, H., Blöch, C., Turner, B., Stuessy, T. F., Obermayer, R., Villaseñor, J. L., & Schneeweiss, G. M. (2012). The evolutionary history of the white-rayed species of *Melampodium* (Asteraceae) involved multiple cycles of hybridization and polyploidization. *American Journal of Botany, 99*, 1043–1057.

Rebernig, C. A., Weiss-Schneeweiss, H., Schneeweiss, G. M., Schönswetter, P., Obermayer, R., Villaseñor, J. L., & Stuessy, T. F. (2010b). Quaternary range dynamics and polyploid evolution in an arid brushland plant species (*Melampodium cinereum*, Asteraceae). *Molecular Phylogenetics and Evolution, 54*, 594–606.

Reich, P. B., Tilman, D., Isbell, F., Mueller, K., Hobbie, S. E., Flynn, D. F. B., & Eisenhauer, N. (2012). Impacts of biodiversity loss escalate through time as redundancy fades. *Science, 336*, 589–592.

Renny-Byfield, S., Kovařík, A., Kelly, L. J., Macas, J., Novák, P., Chase, M. W., Nichols, R. A., Pancholi, M. R., Grandbastien, M.-A., & Leitch, A. R. (2013). Diploidization and genome size change in allopolyploids is associated with differential dynamics of low- and high-copy sequences. *The Plant Journal, 74*, 829–839.

Riddle, B. R., & Hafner, D. J. (2006). A step-wise approach to integrating phylogeographic and phylogenetic biogeographic perspectives on the history of a core North American warm desert biota. *Journal of Arid Environments, 66*, 435–461.

Rieseberg, L. H. (2001). Chromosome rearrangements and speciation. *Trends in Ecology & Evolution, 16*, 351–358.

Rieseberg, L. H., Church, S. A., & Morjan, C. L. (2003). Integration of populations and differentiation of species. *The New Phytologist, 161*, 59–69.

Rieseberg, L. H., & Willis, J. H. (2007). Plant speciation. *Science, 317*, 910–914.

Robinson, B. L. (1901). Synopsis of the genus *Melampodium*. *Proceedings of the American Academy of Arts and Sciences, 36*, 455–466.

Robinson, B. L. (1902). Flora of the Galápagos Islands. *Proceedings of the American Academy of Arts and Sciences, 38*, 208. pl. 1, f. 3.

Robinson, H. (1981). A revision of the tribal and subtribal limits of the Heliantheae (Asteraceae). *Smithsonian Contributions to Botany, 51*, 1–102.

Rzedowski, J. (1978). *Vegetación de México*. Editorial Limusa: México, D.F.

Samuel, R., Gutermann, W., Stuessy, T. F., Ruas, C. F., Lack, H. W., Tremetsberger, K., Talavera, S., Hermanowski, B., & Ehrendorfer, F. (2006). Molecular phylogenetics reveals *Leontodon* (Asteraceae, Lactuceae) to be diphyletic. *American Journal of Botany, 93*, 1193–1205.

Samways, M. J. (1999). Translocating fauna to foreign lands; here comes the Homogenocene. *Journal of Insect Conservation, 3*, 65–66.
Santa María-Díaz, A. (2016). Aspectos geológicos de la Sierra Madre del Sur. In I. Luna-Vega, D. Espinosa, & R. Contreras Medina (Eds.), *Biodiversidad de la Sierra Madre del Sur: Una síntesis preliminar* (pp. 39–66). Universidad Nacional Autónoma de México: Ciudad de México.
Schlüter, P. M., Ruas, P. M., Kohl, G., Ruas, C. F., Stuessy, T. F., & Paulus, H. F. (2009). Genetic patterns and pollination in *Ophrys iricolor* and *O. mesaritica* (Orchidaceae): Sympatric evolution by pollinator shift. *Botanical Journal of the Linnean Society, 159*, 583–598.
Schubert, I. (2007). Chromosome evolution. *Current Opinion in Plant Biology, 10*, 109–115.
Schubert, I., & Vu, G. T. H. (2016). Genome stability and evolution: Attempting a holistic view. *Trends in Plant Science, 21*, 749–757.
Scopece, G., Widmer, A., & Cozzolino, S. (2008). Evolution of postzygotic reproductive isolation in a guild of deceptive orchids. *American Naturalist, 171*, 315–326.
Seaman, F. C. (1982). Sesquiterpene lactones as taxonomic characters in the Asteraceae. *The Botanical Review, 48*, 121–595.
Seaman, F. C., Fischer, N. H., & Stuessy, T. F. (1980). Systematic implications of sesquiterpene lactones in the subtribe Melampodiinae. *Biochemical Systematics and Ecology, 8*, 263–271.
Sedeek, K. E. M., Scopece, G., Staedler, Y. M., Schönenberger, J., Cozzolino, S., Schiestl, F. P., & Schlüter, P. M. (2014). Genic rather than genome-wide differences between sexually deceptive *Ophrys* orchids with different pollinators. *Molecular Ecology, 23*, 6192–6205.
Semple, J. C., & Watanabe, K. (2009). A review of chromosome numbers in Asteraceae with hypotheses on chromosomal base number evolution. In V. A. Funk, A. Susanna, T. F. Stuessy, & R. J. Bayer (Eds.), *Systematics, evolution, and biogeography of Compositae* (pp. 61–72). International Association for Plant Taxonomy: Vienna.
Senderowicz, M., Nowak, T., Rojek-Jelonek, M., Bisaga, M., Papp, L., Weiss-Schneeweiss, H., & Kolano, B. (2021). Descending dysploidy and bidirectional changes in genome size accompanied *Crepis* (Asteraceae) evolution. *Genes, 12*, 1436. https://doi.org/10.3390/genes12091436
Smadja, C. M., & Butlin, R. K. (2011). A framework for comparing processes of speciation in the presence of gene flow. *Molecular Ecology, 20*, 5123–5140.
Sneath, P. H. A., & Sokal, R. R. (1973). *Numerical taxonomy: Principles and practice of numerical classification*. Freeman: San Francisco.
Sokal, R. R., & Rohlf, F. J. (1962). The comparison of dendrograms by objective methods. *Taxon, 11*, 33–40.
Sokal, R. R., & Sneath, P. H. A. (1963). *Principles of numerical taxonomy*. Freeman: San Francisco.
Soltis, D. E., Buggs, R. J. A., Doyle, J. J., & Soltis, P. S. (2010). What we still don't know about polyploidy. *Taxon, 59*, 1387–1403.
Soltis, D. E., Soltis, P. S., & Doyle, J. J. (Eds.). (1998). *Molecular systematics of plants II*. Kluwer: Boston.
Stafleu, F. A., & Cowan, R. S. (1979). *Taxonomic literature. A selective guide to botanical publications and collections with dates, commentaries and types. Volume II: H-Le* (2nd ed.). Bohn, Scheltema & Holkema/W. Junk: The Hague.
Stearn, W. T. (1960). Notes on Linnaeus's 'Genera Plantarum'. In C. Linnaeus (Ed.), *Genera plantarum* (5th ed., pp. v–xxiv). Stockholm. [Facsimile edition, H. R. Engelmann (J. Cramer) and Wheldon & Wesley: Weinheim/Bergstr. and Codicote/Herts.].
Stebbins, G. L. (1950). *Variation and evolution in plants*. Columbia University Press: New York.
Stebbins, G. L. (1971). *Chromosomal evolution in higher plants*. Edward Arnold: London.
Steinbauer, M. J., Field, R., Grytnes, J.-A., Trigas, P., Ah-Peng, C., Attorre, F., Birks, H. J. B., Borges, P. A. V., Cardoso, P., Chou, C.-H., De Sanctis, M., de Sequeira, M. M., Duarte, M. C., Elias, R. B., Fernández-Palacios, J. M., Gabriel, R., Gereau, R. E., Gillespie, R. G., Greimler, J., Harter, D. E. V., Huang, T.-J., Irl, S. D. H., Jeanmonod, D., Jentsch, A., Jump, A. S., Kueffer, C., Nogue, S., Rüdiger, O., Price, J., Romeiras, M. M., Strasberg, D., Stuessy, T., Svenning, J.-C., Vetaas, O. R., & Beierkuhnlein, C. (2016). Topography-driven isolation, speciation and a global increase of endemism with elevation. *Global Ecology and Biogeography, 25*, 1097–1107.

Stevens, W. D., Ulloa, U. C., Pool, A., & Montiel, O. M. (Eds.). (2001). *Flora de Nicaragua. Introducción. Gimnospermas y angiospermas (Acanthaceae—Euphorbiaceae)*. Missouri Botanical Garden Press: St. Louis.

Strother, J. L. (2006). *Melampodium*. In Flora of North America Editorial Committee (Ed.), *Flora of North America, north of Mexico. Vol. 21. Magnoliophyta: Asteridae, part 8: Asteraceae, part 3* (pp. 34–36). Oxford University Press: New York.

Stuessy, T. F. (1968). *A systematic study of the genus Melampodium (Compositae-Heliantheae)*. Ph.D. dissertation. University of Texas: Austin.

Stuessy, T. F. (1969a). Re-establishment of the genus *Unxia* (Compositae-Heliantheae). *Brittonia, 21*, 314–321.

Stuessy, T. F. (1969b). A new variety and new combination in *Melampodium* (Compositae-Heliantheae). *Sida, 3*, 348–349.

Stuessy, T. F. (1970a). Chromosome studies in *Melampodium* (Compositae, Heliantheae). *Madrono, 20*, 365–372.

Stuessy, T. F. (1970b). The genus *Acanthospermum* (Compositae-Heliantheae-Melampodinae): Taxonomic changes and generic affinities. *Rhodora, 72*, 106–109.

Stuessy, T. F. (1970c). Six new species of *Melampodium* (Compositae: Heliantheae) from Mexico and Central America. *Brittonia, 22*, 112–124.

Stuessy, T. F. (1971a). Chromosome numbers and phylogeny in *Melampodium* (Compositae). *American Journal of Botany, 58*, 732–736.

Stuessy, T. F. (1971b). Systematic relationships in the white-rayed species of *Melampodium* (Compositae). *Brittonia, 23*, 177–190.

Stuessy, T. F. (1972). Revision of the genus *Melampodium* (Compositae: Heliantheae). *Rhodora, 74*, 1–70, 161–219.

Stuessy, T. F. (1973a). A systematic review of the subtribe Melampodiinae (Compositae, Heliantheae). *Contributions from the Gray Herbarium, 203*, 65–80.

Stuessy, T. F. (1973b). Revision of the genus *Baltimora* (Compositae, Heliantheae). *Fieldiana Botany, 36*, 31–50.

Stuessy, T. F. (1977). Heliantheae—Systematic review. In V. H. Heywood, J. B. Harborne, & B. L. Turner (Eds.), *The biology and chemistry of the Compositae* (pp. 621–671). Academic Press: London.

Stuessy, T. F. (1978). Revision of *Lagascea* (Compositae, Heliantheae). *Fieldiana Botany, 38*, 75–133.

Stuessy, T. F. (1979). Cladistics of *Melampodium* (Compositae). *Taxon, 28*, 179–195.

Stuessy, T. F. (1980). Supplement to the systematics of the Heliantheae (Compositae). *Boletin de la Sociedad Argentina de Botanica, 19*, 25–32.

Stuessy, T. F. (1993). The role of creative monography in the biodiversity crisis. *Taxon, 42*, 313–321.

Stuessy, T. F. (2009). *Plant taxonomy: The systematic evaluation of comparative data* (2nd ed.). Columbia University Press: New York.

Stuessy, T. F. (2020a). Challenges facing systematic biology. *Taxon, 69*, 655–667.

Stuessy, T. F. (2020b). Opportunities and challenges for research in systematic and evolutionary botany in Latin America. *Gayana Botanica, 77*, 1–10.

Stuessy, T. F. (2025). *Organizing the green world: A conceptual history of botanical classification*. Springer Nature: Cham.

Stuessy, T. F., Blöch, C., Villaseñor, J. L., Rebernig, C. A., & Weiss-Schneeweiss, H. (2011). Phylogenetic analyses of DNA sequences with chromosomal and morphological data confirm and refine sectional and series classification within *Melampodium* (Asteraceae, Millerieae). *Taxon, 60*, 436–449.

Stuessy, T. F., & Brunken, J. N. (1979). Artificial interspecific hybridization in *Melampodium* section *Zarabellia* (Compositae). *Madroño, 26*, 53–63.

Stuessy, T. F., Crawford, D. J., Soltis, D. E., & Soltis, P. S. (2014). *Plant systematics: The origin, interpretation, and ordering of plant biodiversity*. Koeltz Scientific Books: Königstein.

Stuessy, T. F., & Crisci, J. V. (1984). Phenetics of *Melampodium* (Compositae, Heliantheae). *Madroño, 31*, 8–19.
Stuessy, T. F., & Garver, D. (1996). The defensive role of pappus in heads of Compositae. In P. D. S. Caligari & D. J. N. Hind (Eds.), *Compositae: Biology & utilization. Proceedings of the international Compositae conference, Kew, 1994* (Vol. 2, pp. 81–91). Royal Botanic Gardens: Kew.
Stuessy, T. F., & Lack, H. W. (Eds.). (2011). *Monographic plant systematics: Fundamental assessment of plant biodiversity*. A. R. G. Gantner Verlag: Ruggell.
Stuessy, T. F., & Liu, H.-Y. (1983). Anatomy of the pericarp of *Clibadium*, *Desmanthodium* and *Ichthyothere* (Compositae, Heliantheae) and systematic implications. *Rhodora, 85*, 213–227.
Stuessy, T. F., & Spooner, D. M. (1988). The adaptive value and phylogenetic significance of receptacular bracts in the Compositae. *Taxon, 37*, 114–126.
Stuessy, T. F., & Urtubey, E. (2006). Phylogenetic implications of corolla morphology in subfamily Barnadesioideae (Asteraceae). *Flora, 201*, 340–352.
Stuessy, T. F., & Weiss-Schneeweiss, H. (2019). What drives polyploidization in plants? *New Phytologist, 223*, 1690–1692.
Stuessy, T. F., Weiss-Schneeweiss, H., & Keil, D. J. (2004). Diploid and polyploid cytotype distribution in *Melampodium cinereum* and *M. leucanthum* (Asteraceae, Heliantheae). *American Journal of Botany, 91*, 889–898.
Sundberg, S. D., & Stuessy, T. F. (1990). Isolating mechanisms and implications for modes of speciation in Heliantheae (Compositae). In T. J. Mabry & G. Wagenitz (Eds.), *Research advances in the Compositae* (pp. 77–97). Springer: Vienna.
Tiagi, Y. D., & Manilal, K. S. (1964). The morphology and anatomy of the capitulum of *Acanthospermum hispidum* DC. *Proceedings of the National Academy of Sciences, India, 34*, 291–305.
Tilman, D., Reich, P. B., & Knops, J. M. H. (2006). Biodiversity and ecosystem stability in a decade-long grassland experiment. *Nature, 441*, 629–632.
Turner, B. L. (1988). A new species of *Melampodium* (Asteraceae-Heliantheae) from Oaxaca, Mexico. *Phytologia, 64*, 445–447.
Turner, B. L. (1993). A new species of *Melampodium* (Asteraceae, Heliantheae) from Jalisco, Mexico. *Phytologia, 75*, 136–139.
Turner, B. L. (2007). *Melampodium moctezumum* (Asteraceae: Heliantheae), a new species from Sonora, Mexico. *Phytologia, 89*, 258–262.
Turner, B. L. (2016). *Melampodium elottianum* (Asteraceae: Heliantheae): A new species from along the Rio Cuixmala, Jalisco. *Phytologia, 98*, 26–29.
Turner, B. L. (2017). Chapter 23 Tribe Millerieae Lindl. subtribe Melampodiinae. In The Comps (Compositae) of Mexico. A systematic account of the family Asteraceae. *Phytologia Memoirs,* 26b, 13–32, maps 8–22.
Turner, B. L., Beaman, J. H., & Rock, H. F. (1961). Chromosome numbers in the Compositae. V. Mexican and Guatemalan species. *Rhodora, 63*, 121–129.
Turner, B. L., & Flyr, D. (1966). Chromosome numbers in the Compositae. X. North American species. *American Journal of Botany, 53*, 24–33.
Turner, B. L., & Johnston, M. C. (1961). Chromosome numbers in the Compositae-III. Certain Mexican species. *Brittonia, 13*, 64–69.
Turner, B. L., & King, R. M. (1962). A cytotaxonomic survey of *Melampodium* (Compositae-Heliantheae). *American Journal of Botany, 49*, 263–269.
Turner, B. L., & King, R. M. (1964). Chromosome numbers in the Compositae. VIII. Mexican and central American species. *The Southwestern Naturalist, 9*, 27–39.
Tye, A., & Jaramillo, D. P. (2022). Rediscovery of the Galápagos endemic *Lecocarpus leptolobus* (Asteraceae), its morphology, distribution and taxonomy relative to its congeners. *Botanical Journal of the Linnean Society, 200*, 270–284.

Uher, J. (2012). *Sanvitalia speciosa* in the horticultural trade: Unknown origin, uncertain identity but no *Sanvitalia*. *Acta Universitatis Agriculturae et Silviculturae Mendelianae Brunensis, 60*, 339–342.

Villaseñor, J. L., Calzada, J. I., & Dávila, P. (2011). *Melampodium sinuatum* (Asteraceae) endémica del extremo sur de la peninsula de Baja California en peligro de extinción. *Revista Mexicana de Biodiversidad, 82*, 1029–1031.

Villaseñor, J. L., & Ortiz, E. (2025). Floristic richness comparison among the Mexican states. *Revista Mexicana de Biodiversidad, 96*, e965505.

Vitecek, S., Kučinić, M., Previšić, A., Živić, I., Stojanović, K., Keresztes, L., Bálint, M., Hoppeler, F., Waringer, J., Graf, W., & Pauls, S. U. (2017). Integrative taxonomy by molecular species delimitation: Multi-locus data corroborate a new species of Balkin Drusinae micro-endemics. *BMC Evolutionary Biology, 17*, 129. https://doi.org/10.1186/s12862-017-0972-5

Wagner, F., Ott, T., Zimmer, C., Reichhart, V., Vogt, R., & Oberprieler, C. (2019). 'At the cross-roads towards polyploidy': Genomic divergence and extent of homoploid hybridization are drivers for the formation of the ox-eye daisy polyploid complex (*Leucanthemum*, Compositae-Anthemideae). *New Phytologist, 223*, 2039–2053.

Wagner, W. H., Jr. (1961). Problems in the classification of ferns. In D. L. Bailey (Ed.), *Recent advances in botany, from lectures & symposia presented to the IX international botanical congress Montreal 1959* (Vol. 1, pp. 841–844). University of Toronto Press: Toronto.

Wäldchen, J., Rzanny, M., Seeland, M., & Mäder, P. (2018). Automated plant species identification—Trends and future directions. *PLoS Computational Biology, 14*(4), e1005993.

Wang, S., & Loreau, M. (2016). Biodiversity and ecosystem stability across scales in metacommunities. *Ecology Letters, 19*, 510–518.

Watanabe, K. (2008). *Index to chromosome numbers in Asteraceae*. http://www.lib.kobe-u.ac.jp/products/asteraceae/index.html. Accessed 2 Sept 2022.

Watson, J. D., & Crick, F. H. C. (1953). Molecular structure of nucleic acids: A structure for deoxyribose nucleic acid. *Nature, 171*, 737–738.

Weiss-Schneeweiss, H., Blöch, C., Turner, B., Villaseñor, J. L., Stuessy, T. F., & Schneeweiss, G. M. (2012). The promiscuous and the chaste: Frequent allopolyploid speciation and its genomic consequences in American daisies (*Melampodium* sect. *Melampodium*; Asteraceae). *Evolution, 66*, 211–228.

Weiss-Schneeweiss, H., Greilhuber, J., & Schneeweiss, G. M. (2006). Genome size evolution in holoparasitic *Orobanche* (Orobanchaceae) and related genera. *American Journal of Botany, 93*, 148–156.

Weiss-Schneeweiss, H., Leitch, A. R., McCann, J., Jang, T.-S., & Macas, J. (2015). Employing next-generation sequencing to explore the repeat landscape of the plant genome. In E. Hörandl & M. S. Appelhans (Eds.), *Next-generation sequencing in plant systematics* (pp. 155–179). Koeltz Scientific Books: Königstein.

Weiss-Schneeweiss, H., Stuessy, T. F., & Villaseñor, J. L. (2009). Chromosome numbers, karyotypes, and evolution in *Melampodium* (Asteraceae). *International Journal of Plant Sciences, 170*, 1168–1182.

Wen, J., Ickert-Bond, S. M., Appelhans, M. S., Dorr, L. J., & Funk, V. A. (2015). Collections-based systematics: Opportunities and outlook for 2050. *Journal of Systematics and Evolution, 53*, 477–488.

Whiffin, T., & Bierner, M. W. (1972). A quick method for computing Wagner trees. *Taxon, 21*, 83–90.

Wicker, T., Sabot, F., Hua-Van, A., Bennetzen, J. L., Capy, P., Chalhoub, B., Flavell, A., Leroy, P., Morgante, M., Panaud, O., Paux, E., SanMiguel, P., & Schulman, A. H. (2007). A unified classification system for eukaryotic transposable elements. *Nature Reviews Genetics, 8*, 973–982.

Wiggins, I. L., & Porter, D. M. (1971). *Flora of the Galápagos Islands*. Stanford University Press: Stanford.

Wilkins, J. S. (2009a). *Defining species: A sourcebook from antiquity to today*. Peter Lang: New York.

Wilkins, J. S. (2009b). *Species: A history of the idea*. University of California Press: Berkeley.

Wilkins, J. S. (2023). *Understanding species*. Cambridge University Press: Cambridge.

Wood, T. E., Takebayashi, N., Barker, M. S., Mayrose, I., Greenspoon, P. B., & Rieseberg, L. H. (2009). The frequency of polyploid speciation in vascular plants. *Proceedings of the National Academy of Sciences of the United States of America, 106*, 13875–13879.

Yalcin, S., & Leroux, S. J. (2017). Diversity and suitability of existing methods and metrics for quantifying species range shifts. *Global Ecology and Biogeography, 26*, 609–624.

Yigit, E., Sabanci, K., Toktas, A., & Kayabasi, A. (2019). A study on visual features of leaves in plant identification using artificial intelligence techniques. *Computers and Electronics in Agriculture, 156*, 369–377.

Yoshioka, H., Mabry, T. J., & Timmermann, B. N. (1973). *Sesquiterpene lactones: Chemistry, NMR and plant distribution*. University of Tokyo Press: Tokyo.

Zachos, F. E. (2016). *Species concepts in biology: Historical development, theoretical foundations and practical relevance*. Springer: Basel.

Zuckerkandl, E., & Pauling, L. (1965). Molecules as documents of evolutionary history. *Journal of Theoretical Biology, 8*, 357–366.

Index

T. F. Stuessy et al., *Systematics, Ecology, and Chromosomal Evolution of* Melampodium *(Asteraceae)*, https://doi.org/10.1007/978-3-031-81995-7

MIX
Papier aus verantwortungsvollen Quellen
Paper from responsible sources
FSC® C105338

If you have any concerns about our products,
you can contact us on
ProductSafety@springernature.com

In case Publisher is established outside the EU,
the EU authorized representative is:
Springer Nature Customer Service Center GmbH
Europaplatz 3, 69115 Heidelberg, Germany

Printed by Libri Plureos GmbH
in Hamburg, Germany